MATH
at Work

2

Introduction to Algebra

Robert B. Angus
Senior Lecturer
Northeastern University

Claudia A. Clark
Lecturer
Northeastern University

Prentice Hall
Upper Saddle River, New Jersey • Columbus, Ohio

Library of Congress Cataloging-in-Publication Data

Angus, Robert B. (Robert Brownell)
    Math at work / Robert B. Angus, Claudia A. Clark.
        p.   cm.
    "Earlier editions entitled Math is a four letter word" — T.p.
verso.
    Includes index.
    Contents: bk. 1. Review of arithmetic — bk. 2. Introduction to
algebra.
    ISBN 0-13-860388-X (bk. 1 : paper). — ISBN 0-13-857442-1 (bk. 2 :
paper)
        1. Arithmetic.   I. Clark, Claudia A.   II. Title.
QA107.A54   1999   97-45052
513.2—DC21   CIP

Cover art: ©Brian Deep
Editor: Stephen Helba
Production Editor: Louise N. Sette
Copy Editor: Margaret A. Diehl
Design Coordinator: Karrie M. Converse
Text Designer: Rebecca M. Bobb
Cover Designer: Brian Deep
Production Manager: Deidra M. Schwartz
Illustrations: York Graphic Services, Inc.
Marketing Manager: Frank Mortimer

This book was set in Times Roman by York Graphic Services, Inc. and was printed and bound by Banta Company. The cover was printed by Phoenix Color Corp.

 © 1999 by Prentice-Hall, Inc.
Simon & Schuster/A Viacom Company
Upper Saddle River, New Jersey 07458

Earlier editions, entitled *Math Is a Four Letter Word*, ©1997, 1996, 1995, 1990, 1988, 1987, 1985 by Bowen's Publishing.

Printed in the United States of America

10 9 8 7 6 5 4 3 2 1

ISBN: 0-13-857442-1

Prentice-Hall International (UK) Limited, *London*
Prentice-Hall of Australia Pty. Limited, *Sydney*
Prentice-Hall of Canada, Inc., *Toronto*
Prentice-Hall Hispanoamericana, S. A., *Mexico*
Prentice-Hall of India Private Limited, *New Delhi*
Prentice-Hall of Japan, Inc., *Tokyo*
Simon & Schuster Asia Pte. Ltd., *Singapore*
Editora Prentice-Hall do Brasil, Ltda., *Rio de Janeiro*

We dedicate this book to Bob's first-born son

Scott Jeffrey Angus

and to all other first-born children through whom parents learned to be parents

And to Claudia's teachers, friends, and family,
who helped her transform her possibilities into realities

About the Authors

Robert Angus earned a B.S. degree in Electrical Engineering from Northeastern University and an M.S. in Engineering Science from Harvard University. He has taught math, physics, and electrical engineering topics at five colleges and universities in the greater Boston area. After several years in industrial and government work, he founded Bowen's Publishing and now guides the writing of unusual technical books. In addition to coauthoring *Math at Work,* he has written several other books and manuals.

Claudia Clark began work in the human-services and high-tech industries following completion of a B.A. degree from Smith College. She became a tutor for "math-rebellious" students in the greater Boston area. Later, she joined the team of writers and editors at Bowen's Publishing. She has taught precollege and college math to students at Northeastern University and now coauthors math books.

Preface

Math at Work: Introduction to Algebra is the second of a two-book series. These books begin with elementary arithmetic, take you through the fundamentals of arithmetic and introductory algebra, and prepare you for college-level algebra. You can use the books individually, in groups, or with a tutor.

Each chapter explains a math idea. Throughout each chapter there are both Examples, including their solutions or explanations, and numbered Exercise Sets for you to solve on your own. The answers to the Exercise Sets are given in Appendix A. At the end of each chapter there is a practice Chapter Test. The answers to these Chapter Tests, along with self-scoring information, are given in Appendix B.

Each book is divided into numbered parts. Book 1 has Parts 1–4, and Book 2 has Parts 5–8. At the end of every part there is a Part Review Test, similar to a mid-term test. The answers to the Part Review Tests are given in Appendix B. If you have difficulty solving a given exercise, the answer in Appendix B directs you to the page(s) in the book that you need to review.

At the end of each book there is a Book Test, similar to a final exam. The answers to the Book Test are also given in Appendix B. Again, if you have difficulty with a certain problem, the answer in Appendix B directs you to the page(s) that you need to review.

We have tested these books with many students. Some students have worked alone; others have been guided by a teacher or tutor. All of them were preparing for college algebra. Wherever these students experienced difficulty, we included more explanations and practice exercises. Where necessary, we presented the same concept in more than one way. For instance, charts and tables are gradually introduced to show other ways of providing information. Also, the original dictionary pronounciation key in the Glossary of Math Words at the end of each chapter was very difficult for many students to absorb along with the math. Thus, we replaced the key with simpler phonetic pronounciations.

Our students told us two important facts about their use of these books:

    1. Those who prepared 3 x 5 cards (see Appendix D) and reviewed them at least daily (at first) achieved better results than most of their classmates did.

2. Those who had regular study periods, each limited to a total of 15 to 90 quality minutes, retained their knowledge better than those who "crammed."

Appendix C discusses overcoming math anxiety. Follow its general guidelines; it will help you become less anxious while studying the ideas explained in this book.

Appendix D outlines memory methods that have proved successful for others like you.

Appendix E discusses test-taking techniques.

Appendix F contains the prime numbers between 1 and 500. They will become useful to you starting with Chapter 6 and continuing throughout the remainder of these books and your later math studies.

Appendix G (in Book 2 only) describes our recommendations for your purchase of a hand-held calculator.

We believe that these books will be valuable to you in your math studies, and we wish you well!

Robert Angus
Claudia Clark

> *English may be a difficult or second language for you.*
> *The words in this book have been*
> *carefully chosen to ease the reading.*

Acknowledgments

Sheila Tobias, in her book *Math Anxiety*, discussed the problems that have caused people difficulties in learning math. Many of her subjects suffered from the belief that they could never learn math, a negative attitude reinforced by parents, teachers, and counselors. Other problems ranged from information overload to a lack of consistency in the approach to teaching math. We decided to examine these difficulties and provide two books that address them.

Our tests at the University of Rhode Island (Providence) and Northeastern University (Boston, Dedham, and Burlington) have made us realize that our work is successful. We now offer these books to you in the belief that they will assist you in becoming more successful in your pursuit of additional education and learning.

Many persons have contributed to the preparation of this book. Jackie Platt was encouraging when we examined Ms. Tobias's material. Phil Dunphy, Dave Russell, and Willard Whittemore reviewed the contents of each chapter and offered many helpful suggestions. Paula Vosburgh was a continuing supporter during the testing of the two books.

Janice Svendsen is the one person who worked with us at the beginning to develop these two books. She was persistent and full of ideas on how to help others conquer the math they did not learn in school. She has now successfully met her business degree math requirements and is pursuing a career in business.

Natalie Wicks is responsible for the initial design and layout of our text. She contributed many ideas that have also simplified its presentation. An immense thanks is due her. Sheila Keady-Buttaro edited and greatly simplified the wording of all of our initial drafts. She deserves much credit for this effort. Judith Nowinski reedited the latest drafts; introduced chapter subdivisions; and greatly improved the wording, layout, and presentations of the books. We thank her especially for her work. Barbara Coleman spent countless hours cheerfully assisting us in proofreading every word and diagram at least three times. We thank her for her patience and accuracy.

We would also like to thank the reviewers of this book: James R. Boyd, Franklin Institute of Boston; Ky Davis, Muskingum Area Technical College; Marcia Kemen, Wentworth Institute of Technology; Carol Thedford, New Hampshire Community Technical College; and Tim Woo, Durham Technical College.

We realize that some of you may prefer to have a local person answer your questions. Our material is presented in a carefully developed sequence that allows you to explore each idea either with or without assistance. If you do have someone assist you, note that we have usually selected only one method of solution so that you will not have to learn several methods. Therefore, we suggest that you warn any assisting person not to offer you too many additional methods for solving a given type of problem.

Bob Angus
Claudia Clark

To the Instructor

We have written this series of books primarily for students interested in learning math as an introductory or support subject for a college program in science, engineering, technology, business, or liberal arts. We first analyzed college math texts and studied courses to determine the minimum amount of algebra, trigonometry, calculus, and statistics required as a foundation for college courses. Next we changed the traditional sequence of math topics to ease the introduction of new information. Thus, the sequence is different from that used historically in most math texts.

Most non–math majors have difficulty remembering details regarding math procedures. Therefore, we omitted those "tricks" used to perform a computation that do not explain why it is possible and where it cannot be applied. We advise you to avoid offering additional methods for solving a problem until you have used the entire book at least once. If you have very strong feelings about using a different approach, please contact us. We may wish to contact you and discuss your suggestion with you. If we adopt it, then we will credit you in the next edition.

These books were **not** developed for the use of math majors. However, teachers of math may find our approach useful in tutoring math students who are having difficulty with a specific math topic. These books will still prove very helpful in preparing them for college-level algebra, at the very least. Perhaps these students may **later** decide to become math majors!

We have prepared an Instructor's Resource Manual to support these texts. Please contact your local Prentice-Hall representative to obtain a copy.

Any suggestions regarding content, sequence, or examples used should be directed to:

Math Editor
Bowen's Publishing
P.O. Box 270
Bedford, MA 01730-0270

Brief Contents of *Book 1: Review of Arithmetic*

Contents of *Book 2: Introduction to Algebra*

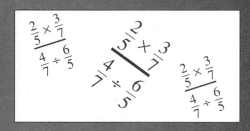

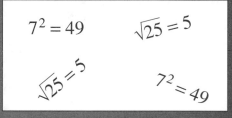

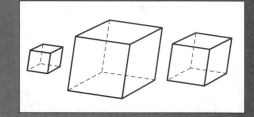

More Fractions

Complex Fractions

Recall that a **simple fraction** is the ratio of two numbers. In this chapter, we will discuss more complicated fractions. Review the fraction simplification procedures from Chapter 9 in Book 1. These procedures will be extended to include the fractions discussed in this chapter.

The procedures for manipulating all of the types of fractions normally found in math courses are the same. They require applying the approach used in Chapter 9 of Book 1.

A fraction may consist of fractions within fractions. Such a fraction is known as a **complex fraction** or a **complicated fraction.** These fractions, like all other fractions, are single-term expressions. Also note that the processing procedures do not change. However, the work necessary to simplify complex fractions will require more time and patience.

There are three fundamental types of complex fractions that we shall now examine:

1. Fractions that contain single fractions in their numerator and their denominator.
2. Fractions that contain the product or quotient of other functions in their numerator and their denominator.
3. Fractions that contain the sum or difference of other fractions in their numerator and perhaps their denominator. They contain more than one inner term.

Simple Fractions within Fractions

The first type of complex fractions are fractions that contain single fractions in their numerator and their denominator.

Study the following examples.

Example

$$\frac{2/3}{6/7}$$

$$= \frac{2}{3} \div \frac{6}{7}$$

$$= \frac{2}{3} \times \frac{7}{6}$$

$$= \frac{2 \times 7}{3 \times 6}$$

$$= \frac{\cancel{2} \times 7}{3 \times \cancel{2} \times 3}$$

$$= \frac{7}{3 \times 3}$$

$$= \frac{7}{9}$$

Note that this is a review of fraction division which we performed in Chapter 9.

Example

$$\frac{8/5}{15/16}$$

$$= \frac{8}{5} \div \frac{15}{16}$$

$$= \frac{8}{5} \times \frac{16}{15}$$

$$= \frac{8 \times 16}{5 \times 15}$$

$$= \frac{128}{75}$$

As noted in Chapter 9 of Book 1, the result is an improper fraction; we can leave it in its present form or convert it to a decimal fraction:

$$= 1.706\ldots$$

Exercise Set 15.1

The answers to the exercises in this chapter are given in Appendix A.

Simplify the following fractions.

1. $\dfrac{1/5}{3/4}$

2. $\dfrac{3/8}{1/4}$

3. $\dfrac{1/6}{3/4}$

4. $\dfrac{5/12}{5/18}$

5. $\dfrac{5/12}{1/18}$

6. $\dfrac{5/18}{7/24}$

Products and Quotients within Fractions

The second type of complex fraction is a fraction that contains the product or quotient of other fractions in its numerator and its denominator. Recall that a fraction is a single-term expression.

Study the next examples.

Example   Simplify the following fraction:

$$\dfrac{\dfrac{2}{3} \div \dfrac{3}{5}}{\dfrac{4}{7} \times \dfrac{5}{6}}$$

Answer

Step 1:   Eliminate the internal division operation first.

$$= \dfrac{\dfrac{2}{3} \times \dfrac{5}{3}}{\dfrac{4}{7} \times \dfrac{5}{6}}$$

Step 2:   Convert the fraction line to division; use the division symbol.

$$= \left(\dfrac{2}{3} \times \dfrac{5}{3}\right) \div \left(\dfrac{4}{7} \times \dfrac{5}{6}\right)$$

Step 3:   Convert all internal fractions to multiplication of fractions.

$$= \dfrac{2}{3} \times \dfrac{5}{3} \times \dfrac{7}{4} \times \dfrac{6}{5}$$

Note that both denominator fractions were inverted.

Step 4:   Combine the four fractions into one fraction; factor all numbers into primes.

$$= \dfrac{2 \times 5 \times 7 \times 2 \times 3}{3 \times 3 \times 2 \times 2 \times 5}$$

Step 5:   Cancel numbers common to both numerator and denominator.

$$= \dfrac{\cancel{2} \times \cancel{5} \times 7 \times \cancel{2} \times \cancel{3}}{\cancel{3} \times 3 \times \cancel{2} \times \cancel{2} \times \cancel{5}}$$

Step 6:   The remaining factors are the answer to this problem.

$$= \dfrac{7}{3}$$

Example   Simplify the following fraction:

$$\frac{\dfrac{25}{7} \div \dfrac{8}{3}}{\dfrac{5}{4} \div \dfrac{7}{3}}$$

Answer

$$= \left(\frac{25}{7} \div \frac{8}{3}\right) \div \left(\frac{5}{4} \div \frac{7}{3}\right)$$

$$= \left(\frac{25}{7} \times \frac{3}{8}\right) \div \left(\frac{5}{4} \times \frac{3}{7}\right)$$

$$= \frac{25}{7} \times \frac{3}{8} \times \frac{4}{5} \times \frac{7}{3}$$

$$= \frac{25 \times 3 \times 4 \times 7}{7 \times 8 \times 5 \times 3}$$

$$= \frac{5 \times 5 \times 3 \times 2 \times 2 \times 7}{7 \times 2 \times 2 \times 2 \times 5 \times 3}$$

$$= \frac{5}{2}$$

$$= 2.5$$

Exercise Set 15.2

Simplify the following fractions.

1. $\dfrac{\dfrac{1}{2} \times \dfrac{3}{4}}{\dfrac{3}{8} \times \dfrac{1}{6}}$

2. $\dfrac{\dfrac{1}{2} \times \dfrac{3}{4}}{\dfrac{3}{8} \div \dfrac{1}{6}}$

3. $\dfrac{\dfrac{1}{2} \div \dfrac{3}{4}}{\dfrac{3}{8} \div \dfrac{1}{6}}$

4. $\dfrac{\dfrac{1}{2} \times \dfrac{3}{4}}{\dfrac{3}{8} \div \dfrac{1}{5}}$

MATH AT WORK

Sums and Differences within Fractions

The third type of complex fraction is a fraction that contains the sum or difference of other fractions in its numerator and perhaps also in its denominator. It contains more than one **inner term.**

There are two procedures that can be used to simplify these fractions, as the following example illustrates.

Example   Simplify the following fraction:

$$\frac{\dfrac{3}{4} + \dfrac{5}{6}}{\dfrac{7}{2} - \dfrac{2}{3}}$$

Procedure 1

Step 1:   **Separately** determine the numerator and denominator LCDs.

$$= \frac{\dfrac{3}{2 \times 2} + \dfrac{5}{2 \times 3}}{\dfrac{7}{2} - \dfrac{2}{3}} \qquad \frac{\text{numerator LCD: } 2 \times 2 \times 3}{\text{denominator LCD: } 2 \times 3}$$

Step 2:   Separately combine numerators and denominators.

$$= \frac{\dfrac{3 \times 3 + 5 \times 2}{2 \times 2 \times 3}}{\dfrac{7 \times 3 - 2 \times 2}{2 \times 3}}$$

$$= \frac{\dfrac{9 + 10}{2 \times 2 \times 3}}{\dfrac{21 - 4}{2 \times 3}}$$

$$= \frac{\dfrac{19}{2 \times 2 \times 3}}{\dfrac{17}{2 \times 3}}$$

$$= \frac{19}{2 \times 2 \times 3} \div \frac{17}{2 \times 3}$$

Step 3:   Invert the second fraction and multiply.

$$= \frac{19}{2 \times 2 \times 3} \times \frac{2 \times 3}{17}$$

$$= \frac{19}{2 \times 17}$$

$$= 19/34$$

MORE FRACTIONS

Procedure 2

$$\frac{\dfrac{3}{4} + \dfrac{5}{6}}{\dfrac{7}{2} - \dfrac{2}{3}}$$

Step 1: Determine all four LCDs.

Denominator Factors			
First	2	2	
Second	2		3
Third	2		
Fourth			3
LCD	2	2	3

Step 2: Multiply the numerator and the denominator by this LCD.

$$= \frac{\dfrac{3}{2 \times 2} + \dfrac{5}{2 \times 3}}{\dfrac{7}{2} - \dfrac{2}{3}} \times \frac{2 \times 2 \times 3}{2 \times 2 \times 3}$$

Step 3: Distribute the $2 \times 2 \times 3$ to both inner terms in the numerator and both inner terms in the denominator.

$$= \frac{\dfrac{3}{2 \times 2} \times 2 \times 2 \times 3 + \dfrac{5}{2 \times 3} \times 2 \times 2 \times 3}{\dfrac{7}{2} \times 2 \times 2 \times 3 - \dfrac{2}{3} \times 2 \times 2 \times 3}$$

Step 4: In each inner term, cancel the common factors.

$$= \frac{\dfrac{3}{\cancel{2} \times \cancel{2}} \times \cancel{2} \times \cancel{2} \times 3 + \dfrac{5}{\cancel{2} \times \cancel{3}} \times \cancel{2} \times 2 \times \cancel{3}}{\dfrac{7}{\cancel{2}} \times \cancel{2} \times 2 \times 3 - \dfrac{2}{\cancel{3}} \times 2 \times 2 \times \cancel{3}}$$

Step 5: Simplify the result.

$$= \frac{3 \times 3 + 5 \times 2}{7 \times 2 \times 3 - 2 \times 2 \times 2}$$

$$= \frac{9 + 10}{42 - 8}$$

$$= \frac{19}{34}$$

The following example is worked using both procedures.

Example

$$\frac{\frac{6}{7} - \frac{2}{3}}{\frac{4}{15} + \frac{2}{5}}$$

Answer

Procedure 1

$$\frac{\frac{6}{7} - \frac{2}{3}}{\frac{4}{15} + \frac{2}{5}}$$

$$= \frac{\left(\frac{6}{7} - \frac{2}{3}\right) \times \frac{7 \times 3}{7 \times 3}}{\left(\frac{4}{15} + \frac{2}{5}\right) \times \frac{3 \times 5}{3 \times 5}}$$

$$= \frac{\frac{6 \times 3 - 2 \times 7}{7 \times 3}}{\frac{4 + 2 \times 3}{3 \times 5}}$$

$$= \frac{\frac{18 - 14}{7 \times 3}}{\frac{4 + 6}{3 \times 5}}$$

$$= \frac{\frac{4}{21}}{\frac{10}{15}}$$

$$= \frac{4}{21} \div \frac{10}{15}$$

$$= \frac{4}{21} \times \frac{15}{10}$$

$$= \frac{60}{210}$$

$$= \frac{2}{7}$$

Procedure 2

$$\frac{\dfrac{6}{7} - \dfrac{2}{3}}{\dfrac{4}{15} + \dfrac{2}{5}}$$

$$= \frac{\dfrac{6}{7} - \dfrac{2}{3}}{\dfrac{4}{15} + \dfrac{2}{5}} \times \frac{3 \times 5 \times 7}{3 \times 5 \times 7}$$

$$= \frac{\dfrac{6}{7} \times 3 \times 5 \times 7 - \dfrac{2}{3} \times 3 \times 5 \times 7}{\dfrac{4}{15} \times 3 \times 5 \times 7 + \dfrac{2}{5} \times 3 \times 5 \times 7}$$

$$= \frac{6 \times 3 \times 5 - 2 \times 5 \times 7}{4 \times 7 + 2 \times 3 \times 7}$$

$$= \frac{90 - 70}{28 + 42}$$

$$= \frac{20}{70}$$

$$= \frac{2}{7}$$

Exercise Set 15.3

Simplify and reduce the following fractions.

1. $$\dfrac{\dfrac{3}{5} - \dfrac{1}{3}}{\dfrac{5}{6} + \dfrac{1}{2}}$$

2. $$\dfrac{\dfrac{4}{9} + \dfrac{2}{5}}{\dfrac{2}{15} - \dfrac{1}{3}}$$

3. $$\dfrac{\dfrac{1}{6} + \dfrac{2}{9}}{\dfrac{2}{3} + \dfrac{4}{9}}$$

4. $$\dfrac{\dfrac{1}{14} + \dfrac{3}{7}}{\dfrac{4}{21} - \dfrac{1}{6}}$$

5. $\dfrac{\dfrac{3}{4} - \dfrac{4}{5}}{\dfrac{4}{5} - \dfrac{1}{2}}$

6. $\dfrac{\dfrac{3}{4} \times \dfrac{4}{5}}{\dfrac{4}{5} - \dfrac{1}{2}}$ (This problem requires more thought.)

Summary

Complex fractions contain fractions within fractions. Both the numerator and the denominator must be simplified before other math operations can be performed, as the following example shows.

Example

$$\frac{\dfrac{1}{2} + \dfrac{3}{4}}{\dfrac{1}{2} - \dfrac{1}{3}} = \frac{\dfrac{1}{2} + \dfrac{3}{2 \times 2}}{\dfrac{1}{2} - \dfrac{1}{3}}$$

The LCDs for all four fractions are as follows:

Denominator Factors			
First	2		
Second	2	2	
Third	2		
Fourth			3
LCD	2	2	3

$$= \frac{\dfrac{1}{2} + \dfrac{3}{2 \times 2}}{\dfrac{1}{2} - \dfrac{1}{3}} \times \frac{2 \times 2 \times 3}{2 \times 2 \times 3}$$

$$= \frac{2 \times 3 + 3 \times 3}{2 \times 3 - 2 \times 2}$$

$$= \frac{6 + 9}{6 - 4} = \frac{15}{2}$$

The result is **15/2**. Recall from Chapter 9 of Book 1 that this result is an **improper fraction** because its numerator is larger than its denominator. If desired, it could be converted to a decimal fraction:

$$= 7.5$$

Glossary of Math Words

Complex (kom′-pleks) **fraction**   A fraction within a fraction; also known as a **complicated fraction.**

Complicated fraction   *See* Complex fraction.

Improper (im′-prop-er) **fraction**   A fraction where the numerator value is larger than the denominator value; for example, 10/9 is an improper fraction.

Chapter 15 Test

Follow instructions carefully:

> On a separate piece of paper, write the answers to the following questions. Do not write on these pages.
>
> When you are finished, compare your answers with those given in Appendix B.
>
> Record the date, your test time, and your score on the chart at the end of this test.
>
> *Note:* Because of the complexity of this fraction test, the number of test problems has been limited to ten. Each problem is worth two points rather than one.

Simplify the following fractions.

1. $\dfrac{7/12}{1/8}$

2. $\dfrac{\dfrac{5}{6} \times \dfrac{1}{4}}{\dfrac{7}{2} \div \dfrac{9}{10}}$

3. $\dfrac{\dfrac{3}{5} \div \dfrac{1}{6}}{\dfrac{4}{9} \div \dfrac{2}{15}}$

Combine and then simplify the following fractions.

4. $\dfrac{\dfrac{1}{2} - \dfrac{2}{3}}{\dfrac{3}{4} - \dfrac{1}{6}}$

5. $\dfrac{\dfrac{4}{5} - \dfrac{3}{4}}{\dfrac{5}{6} - \dfrac{1}{4}}$

6. $$\dfrac{\dfrac{5}{12}+\dfrac{3}{20}}{\dfrac{6}{5}-\dfrac{4}{15}}$$

7. $$\dfrac{5/8}{\dfrac{4}{5}-\dfrac{1}{4}}$$

8. Simplify the following fraction:

$$\dfrac{\dfrac{2}{3}\div\dfrac{4}{5}}{\dfrac{1}{4}\times\dfrac{2}{5}}$$

9. Combine and then simplify the following fraction:

$$\dfrac{\dfrac{2}{3}\quad\dfrac{1}{4}}{\dfrac{2}{5}-\dfrac{1}{3}}$$

10. Combine and then simplify the following fraction:

$$\dfrac{\dfrac{2}{3}-\dfrac{1}{4}}{\dfrac{2}{5}+\dfrac{1}{3}}$$

Chapter 15 Test Record

DATE	TIME	SCORE

Refer to

- Appendix B for the correct answers to this test.
- Appendix C if taking this test required too much effort.
- Appendix D for **Memory Methods** assistance.
- Appendix E if your test scores are decreasing.

When a problem seems difficult, find one like it in this chapter. Then study that (and the related material) again.

Develop additional 3 × 5 cards for those ideas, problems, and procedures that caused you difficulty.

Exponents and Radicals

Writing Efficient Expressions

It is important to write mathematics expressions efficiently. For example, when we add together six numbers that are all 4's,

$$4 + 4 + 4 + 4 + 4 + 4 = 24$$

we shorten the writing by using multiplication notation:

$$6 \times 4 = 24$$

Recall from Chapter 4 in Book 1 that **multiplication is iterative addition.** (*Iterative* is the math word for "repeated.")

The inverse operation is also important. How many 4's are there in 24? Four **(4)** can be subtracted from **24** exactly six times:

$$24 - 4 - 4 - 4 - 4 - 4 - 4 = 0 \quad \text{remainder}$$

We shorten the writing by using division notation:

$$24 \div 4 = 6$$

Recall from Chapter 5 in Book 1 that **division is iterative subtraction.**

How Exponents Express Iterative Multiplication

Scientists explore the more complicated features of our universe. They need a notation that efficiently indicates iterative (repeating) multiplication. For example, let's reduce the following writing by using simpler notation:

$$3 \times 3 \times 3 \times 3 \times 3 = 243$$

There are five **3**'s multiplied together. Write the number **5** above and to the right of the number **3**.

$$3 \times 3 \times 3 \times 3 \times 3$$
$$= 3^5$$
$$= 243$$

The 3^5 notation is known as **exponential notation.** (This notation was introduced by René Descartes in 1637.) The **5** indicates the number of **3**'s to be multiplied together. Note that the number **5** is written smaller than the number **3**. The **5** is located in a **superscript** position. Examine the location of the number **7** in this example:

$$5^7$$

The **7** is in a superscript position. When the **7** is located above and to the right of another number, then the **7** is known as that number's **exponent.** It indicates the number of **5**'s to be multiplied together.

$$5 \times 5 \times 5 \times 5 \times 5 \times 5 \times 5$$
$$= 5^7$$
$$= 78\ 125$$

All modern computers and their printers can print superscripts. Another computer printer and programming notation for a superscript is

$$5 \wedge 7$$

Exercise Set 16.1

The answers to the exercises in this chapter are given in Appendix A.

Evaluate the following expressions.

1. 3^6

2. 4^2

3. 10^3

4. 2^9

5. 8^3

6. 1^9

7. 5^4

8. 25^2

9. 2^1

Note in Exercise 9 on the preceding page that 2^1 means **one 2.** There is only one **2;** no other number is involved. Therefore, $2^1 = 2.$ All numbers have a **1** exponent that is **understood** to be there. Unless we are verifying a math law involving exponents, we omit the exponent **1.**

Also note that all of the expressions in Exercise Set 16.1 are single-term expressions. Why? Because multiplication is the only math operation involved.

Examine the expression $4^6.$ It is spoken "**four to the sixth power**," or "**four to the sixth.**" The exponent **6** is also known as a **power** because it is so powerful in the expanding of numbers.

Two special cases are spoken differently because they appear so often. They are the exponents **2** and **3.** Thus, 4^2 is spoken "**four squared**," and 4^3 is spoken "**four cubed.**"

The number **4** is known as the **base** of the two numbers. Our decimal system is said to have a base of **ten** because each column represents a power of ten or a product of tens. For example, the tens column is $10^1 = 10,$ the hundreds column is $10^2 = 100,$ and the thousands column is $10^3 = 1000.$ This, along with other related material, is discussed later in this chapter.

As a summary, note the following:

$$\text{Base} \rightarrow 4^6 \leftarrow \text{Exponent, or power}$$

How would you graphically display $4^4,$ $4^5,$ or 4^6? Because these exponents represent values beyond the third dimension, they cannot be visually displayed.

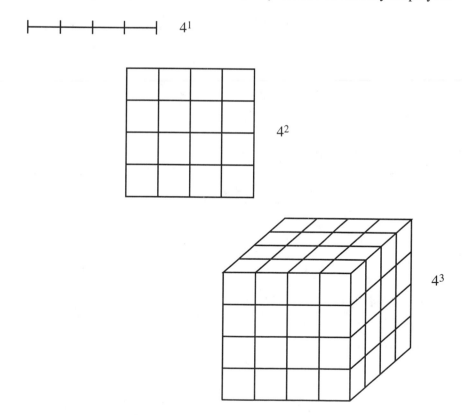

Exercise Set 16.2

Identify the base and the power of each of the following expressions. Then write the longer form, such as $3^2 = 3 \times 3$, and calculate the result, such as $3^2 = 3 \times 3 = 9$.

1. 5^4

2. 21^2

3. 2^{10}

4. 10^2

5. 7^3

6. 3^3

7. 7^5

8. 12^3

9. 2^{20}

10. 10^4

11. 6^4

12. 3^5

MATH AT WORK

Laws of Exponents

There are five rules, or **laws,** that govern exponential operations. We will now discuss each of these five laws and illustrate them with examples.

Law 1 When two expressions that have the same base are multiplied together, then their exponents are added:

$$2^3 \times 2^4$$
$$= (2 \times 2 \times 2) \times (2 \times 2 \times 2 \times 2)$$
$$= 2 \times 2 \times 2 \times 2 \times 2 \times 2 \times 2$$
$$= 2^7$$
$$= 128$$

Thus,

$$2^3 \times 2^4$$
$$= 2^{3+4}$$
$$= 2^7$$
$$= 128$$

Work Exercise Set 16.3 on page 16–11. Then return to this page and continue with the text.

Law 2 When expressions that have the same base are divided, then the exponent of the denominator is subtracted from the exponent of the numerator:

$$\frac{2^5}{2^3}$$
$$= \frac{2 \times 2 \times 2 \times 2 \times 2}{2 \times 2 \times 2}$$
$$= \frac{2 \times 2 \times \cancel{2} \times \cancel{2} \times \cancel{2}}{\cancel{2} \times \cancel{2} \times \cancel{2}}$$

$$= \frac{2 \times 2}{1}$$
$$= 2 \times 2$$
$$= 2^2$$
$$= 4$$

Thus,

$$\frac{2^5}{2^3}$$
$$= 2^{5-3}$$
$$= 2^2$$
$$= 4$$

Work Exercise Set 16.4 on page 16–13. Then go to page 16–10 and continue with the text.

Zero as Exponent   Two special results fascinated early mathematicians who were studying exponents. The first result involved the exponent value **zero,** as the following example illustrates.

Example

$$\frac{2^3}{2^3} \qquad \text{and} \qquad \frac{10^4}{10^4}$$

$$= \frac{2 \times 2 \times 2}{2 \times 2 \times 2} \qquad\qquad = \frac{10 \times 10 \times 10 \times 10}{10 \times 10 \times 10 \times 10}$$

$$= \frac{1}{1} \qquad\qquad\qquad = \frac{1}{1}$$

$$= 1 \qquad\qquad\qquad = 1$$

Thus,

$$2^3 \div 2^3 \qquad \text{and} \qquad 10^4 \div 10^4$$

$$= 2^{3-3} \qquad\qquad = 10^{4-4}$$

$$= 2^0 \qquad\qquad\quad = 10^0$$

$$= 1 \qquad\qquad\quad = 1$$

After studying many other divisions using a variety of bases, these early mathematicians finally agreed that any base **raised** to the zero power would equal **one:**

$$2^0 = 1 \qquad\qquad 7^0 = 1$$
$$5^0 = 1 \qquad\qquad 10^0 = 1$$
$$(95)^0 = 1 \qquad\qquad (7685)^0 = 1$$

Note: 0^0 is undefined; it is the only exception.

Work Exercise Set 16.5 on page 16–15. Then go on to page 16–17 and continue with the text.

Exercise Set 16.3

Work the following exercises by applying the **first law of exponents.**

1. $3^2 \times 3^3 =$

2. $5^0 \times 5^0 =$

3. $7^1 \times 7^2 =$

4. $2^4 \times 2^6 =$

5. $1^3 \times 1^5 =$

6. $4^2 \times 4^3 =$

7. $7^3 \times 7^2 =$

8. $3^1 \times 3^2 =$

9. $2^6 \times 2^7 =$

10. $2^3 \times 1^7 =$

Now return to page 16–9 and continue with the text.

MATH AT WORK

Exercise Set 16.4

Work the following exercises by applying the **second law of exponents.**

1. $\dfrac{3^7}{3^4} =$

2. $4^7 \div 4^6 =$

3. $\dfrac{5^8}{5^6} =$

4. $\dfrac{1^9}{1^6} =$

5. $\dfrac{5^7}{5^5} =$

6. $\dfrac{6^9}{6^4} =$

7. $9^8 \div 9^6 =$

8. $\dfrac{7^6}{7^4} =$

Now return to page 16–10 and continue with the text.

MATH AT WORK

Exercise Set 16.5

Evaluate the following expressions.

1. 9^0

2. $(96)^0$

3. $(86)^0$

4. 1^0

EXPONENTS AND RADICALS

5. $(27)^0$

6. $(-19)^0$

7. $(2\ 037\ 605)^0$

8. $\left(-\dfrac{1}{26}\right)^0$

Now go on to the next page and continue with the text.

The second special result involved the meaning of negative exponents. In the following example, examine the ways in which division involving exponents may be written.

Example

$$2^3 \div 2^5 \qquad\qquad \frac{2^3}{2^5}$$

$$= 2^{3-5} \qquad\qquad = \frac{2 \times 2 \times 2}{2 \times 2 \times 2 \times 2 \times 2}$$

$$= 2^{-2} \qquad\qquad = \frac{1}{2^2}$$

Therefore,

$$2^{-2} = \frac{1}{2^2} \quad \text{which equals } \frac{1}{4} = 0.25$$

Again, after mathematicians studied many possible combinations that resulted in negative exponents, the third law evolved.

Law 3 When a number with an exponent is moved from a numerator to a denominator, change the sign of the exponent. In other words, when determining the reciprocal of any base that has an exponent, change the sign of that exponent. Here are four different examples.

Examples

$$7^4 = \frac{1}{7^{-4}}$$

$$3^{-4} = \frac{1}{3^4}$$

$$\frac{1}{2} = 2^{-1}$$

$$\frac{1}{5^{-2}} = 5^2$$

Thus,

$$2^{-1} = \frac{1}{2} = 0.5$$

You have changed its form, not its value!

Note that the same rule applies when you are moving a number with an exponent from a denominator to a numerator.

Work Exercise Set 16.6 on page 16–19. Then go to page 16–18 and continue with the text.

Previously, you practiced converting negative exponents and positive exponents to the opposite-signed exponents. In the following example, we use this procedure to determine the decimal-number value of two exponential expressions.

Example

$$\frac{1}{5^{-2}} = 5^2 = 25$$

$$2^{-4} = \frac{1}{2^4} = \frac{1}{16} = 0.0625$$

Work Exercise Set 16.7 on page 16–21. Then return to this page and continue with the text.

Law 4   When an expression that consists of a base and an exponent is raised to another exponent (or power), then the exponents are multiplied together:

$$(2^2)^3$$
$$= (2 \times 2)^3$$
$$= (2 \times 2) \times (2 \times 2) \times (2 \times 2)$$
$$= 2 \times 2 \times 2 \times 2 \times 2 \times 2$$
$$= 2^6$$
$$= 64$$

Thus,

$$(2^2)^3$$
$$= 2^{2 \times 3}$$
$$= 2^6$$
$$= 64$$

Work Exercise Set 16.8 on page 16–23. Then go on to page 16–25 and continue with the text.

Exercise Set 16.6

Use the **third law of exponents** to change the sign of each exponent without changing the value of the expression.

1. 2^{-2}

2. 2^2

3. $\dfrac{1}{2^2}$

4. 5^{-4}

5. 7^5

6. $\frac{1}{9}$

7. 1×10^8

8. 1×10^{-6}

9. $\frac{1}{4}$

10. 4^7

Now return to page 16–18 and continue with the text.

Exercise Set 16.7

Determine the decimal fraction for each of the following exercises.

1. 2^{-3}

2. 3^{-2}

3. $\dfrac{5^2}{5^5}$

4. 6^{-2}

5. $\dfrac{7^5}{7^6}$

6. 8^{-1}

7. 2^{-2}

8. 4^{-3}

Now return to page 16–18 and continue with the text.

MATH AT WORK

Exercise Set 16.8

Work the following exercises using the **fourth law of exponents.** Convert all answers to positive exponents.

1. $(3^2)^3 =$

2. $(2^2)^3 =$

3. $(4^2)^3 =$

4. $(5^1)^2 =$

5. $(2^{-2})^3 =$

6. $(2^{-2})^{-3} =$

7. $(3^{-3})^{-3} =$

8. $(7^{-2})^4 =$

9. $(1^{-2})^5 =$

10. $(2^{-2})^0 =$

11. $(10^{-2})^3 =$

12. $(10^{-2})^{-3} =$

Now go on to the next page and continue with the text.

Law 5 When the product or quotient of two or more expressions is raised to a power, then the exponents in each of these expressions are multiplied by that power. The following example illustrates this law.

Example

$$(2^2 \times 3^2)^3$$
$$= (2^2 \times 3^2) \times (2^2 \times 3^2) \times (2^2 \times 3^2)$$
$$= 2 \times 2 \times 3 \times 3 \times 2 \times 2 \times 3 \times 3 \times 2 \times 2 \times 3 \times 3$$
$$= 2^6 \times 3^6$$
$$= 64 \times 729$$
$$= 46\ 656$$

Thus,

$$(2^2 \times 3^2)^3$$
$$2^{2 \times 3} \times 3^{2 \times 3}$$
$$= 2^6 \times 3^6$$
$$= 64 \times 729$$
$$= 46\ 656$$

Once you are able to manipulate expressions involving exponents, you may wish to perform your computations with a calculator. The minimal and pre-ferred keys are given in Appendix G for an algebraic calculator. We do **not** rec-ommend starting with an RPL (Reverse Polish Logic) calculator unless you plan to major in mathematics.

Exercise Set 16.9

Work the following exercises involving the **fifth law of exponents.** Leave the result as either an integer or a decimal fraction.

1. $(2^2 \times 4^2)^3 =$

2. $(3^2 \times 2^2)^3 =$

3. $(2^3 \times 3^2)^3 =$

4. $(5^3 \times 2^3)^4 =$

5. $(2^2 \times 4^3)^2 =$

6. $(3^3 \times 2^{-3})^2 =$

7. $(2^{-2} \times 4^3)^{-2} =$

8. $(3^2 \times 4^{-3})^2 =$

9. $(2^{-2} \times 4^{-3})^{-3} =$

10. $(3^{-2} \times 5^3)^{-2} =$

11. $(4^3 \times 4^{-3})^5 =$

12. $(5^{-3} \times 2^{-5})^{-2} =$

The fifth law also applies to the division of numbers. Work these examples.

13. $(2^3 \div 5^3)^4 =$

14. $(2^3 \div 2^2)^4 =$

15. $(3^2 \div 3^4)^2 =$

16. $(5^3 \div 5^5)^2 =$

17. $(6^3 \div 2^2)^2 =$

18. $(2^{-3} \div 5^3)^{-4} =$

19. $(3^3 \div 2^{-2})^4 =$

20. $(3^{-2} \div 3^{-4})^{-2} =$

21. $(5^{-3} \div 5^5)^{-2} =$

22. $(6^3 \div 2^{-2})^{-2} =$

When the Rules Don't Apply

Do not try to use the rules where they do not apply. For example, numbers cannot be added or subtracted unless their exponents are exactly **one.** Underline each term to help you identify them. For example:

$$\underline{2^3} + \underline{2^2}$$
$$= \underline{8^1} + \underline{4^1}$$
$$= \underline{8} + \underline{4}$$
$$= \underline{12}$$

The laws of exponents apply **only** to the multiplication or division of exponential expressions, that is, only to **one-term** expressions!

Work Exercise Set 16.10 on page 16–33. Then return to this page and continue with the text.

Scientific Notation

Recall from Chapter 1 in Book 1 that our number system is based upon ten decimal symbols:

$$0, \quad 1, \quad 2, \quad 3, \quad 4, \quad 5, \quad 6, \quad 7, \quad 8, \quad 9$$

Numbers greater than nine are written using two or more of these ten decimal symbols side by side, in columns.

In Chapter 10, we introduced decimal fractions. You learned the names of the columns to the right of the decimal points. For all numbers, the column in which a decimal symbol is located determines the value or the worth of the number. For example:

0.3	three-tenths
3	three
30	three tens, or thirty
300	three hundreds, or three hundred

Remember that the placeholder **zero** is very important in number placement.

Let's examine two numbers that require many columns to write.

The number symbols for three trillion: 3 000 000 000 000

The number symbols for two-billionths: 0.000 000 002

Powers of ten, or **scientific notation,** may be used to write such very large and very small numbers more efficiently. This notation combines our knowledge of number placement, exponents, and decimal numbers.

The number **40** can be written as

$$4 \times 10 \quad \text{or} \quad 4 \times 10^1$$

We have written **40** using scientific notation: **4×10^1.**

A number written using scientific notation is the product of

1. a number greater than or equal to one and less than ten (for example, **4**), and
2. ten raised to a power (for example, **10^1**).

The number **2000** can be written as follows using scientific notation:

$$2 \times 1000$$
$$= 2 \times 10 \times 10 \times 10$$
$$= 2 \times 10^3$$

where the **2** is a number from one to ten, and the exponent **3** indicates the number of 10's that are multiplied together.

Therefore,

3 000 000 000 000 can be written as 3×10^{12}, and

0.000 000 002 can be written as 2×10^{-9}.

Here are additional examples.

Example

$$200 = 2 \times 100 = 2 \times 10 \times 10 = 2 \times 10^2$$

Thus, $200 = 2 \times 10^2$.

Example

$$50\ 000 = 5 \times 10\ 000 = 5 \times 10 \times 10 \times 10 \times 10 = 5 \times 10^4$$

Thus, $50\ 000 = 5 \times 10^4$.

Example

$$0.4 = 4 \times 0.1 = 4 \times (1/10) = 4 \times (1/10^1) = 4 \times 10^{-1}$$

Thus, $0.4 = 4 \times 10^{-1}$.

Example

$$0.07 = 7 \times 0.01 = 7 \times (1/100) = 7 \times (1/10^2) = 7 \times 10^{-2}$$

Thus, $0.07 = 7 \times 10^{-2}$.

In each example, the number has been written as the product of a number and ten to a power.

Converting a Decimal Number to Scientific Notation

The following simple two-part rule allows you to quickly determine the scientific notation version of a given number.

Step 1: Examine the given number and locate the **decimal points.**

Step 2: Move the decimal point so that the resulting number is greater than or equal to one and less than ten. As you move the decimal point, note the following:

   a. For each digit that you move the points to the left, add **positive one** to the exponent of ten. For example:

   $$300 = 30.0 \times 10^1 = 3.0 \times 10^2$$

   Thus, $300 = 3 \times 10^2$.

   b. For each digit that you move the points to the right, add **negative one** to the exponent of ten. For example:

   $$0.04 = 0.4 \times 10^{-1} = 4.0 \times 10^{-2}$$

   Thus, $0.04 = 4 \times 10^{-2}$

These rules work for both positive and negative numbers.

Study the following examples.

Examples

$$9000 = 9000. = 900.0 \times 10^1 = 90.00 \times 10^2 = 9.000 \times 10^3$$
$$= 9 \times 10^3$$

$$0.06 = 00.6 \times 10^{-1} = 006. \times 10^{-2}$$
$$= 6 \times 10^{-2}$$

$$260 = 260. = 26.0 \times 10^1 = 2.60 \times 10^2$$
$$= 2.6 \times 10^2$$

$$5100.9 = 510.09 \times 10^1 = 51.009 \times 10^2$$
$$= 5.1009 \times 10^3$$

$$0.47 = 04.7 \times 10^{-1}$$
$$= 4.7 \times 10^{-1}$$

Now return to page 16–29 and continue with the text.

Do Exercise Set 16.11 on page 16–35. Then go on to page 16–37 and continue with the text.

EXPONENTS AND RADICALS

Exercise Set 16.10

Work the following exercises.

1. $2^3 - 2^2 =$

2. $2^2 - 2^3 =$

3. $3^2 + 2^3 =$

4. $3^2 - 2^3 =$

5. $3^2 + 2^3 - 4^2 =$

6. $2^{-3} + 2^3 =$

7. $3^2 + 3^3 =$

8. $3^4 + 2^{-2} - 2^3 =$

9. $5^2 - 2^5 =$

10. $4^{-2} + 2^4 - 5^{-2} =$

Exercise Set 16.11

Write the following numbers using scientific notation. Recall that the left-most factor must be a number from one to ten.

1. 0.000 02 =

2. 4720 =

3. 0.035 =

4. 60 300 =

5. 0.000 000 27 =

6. 245 000 000 =

7. −27 000 =

8. 0.000 306 =

9. −0.0829 =

10. 52 728 =

11. 0.000 092 716 =

Now go on to the next page and continue with the text.

Converting from Scientific Notation to a Decimal Number

How can you convert from scientific notation to a decimal number? Reverse the process, as shown in the following examples.

Examples

$$2.6 \times 10^3 = 2.6 \times 10 \times 10 \times 10 = 2600$$

$$4.7 \times 10^{-2} = 4.7 \times \frac{1}{10^2} = 4.7 \times \frac{1}{10} \times \frac{1}{10} = 0.047$$

You use the simple two-part rule in reverse to convert the given number to a decimal number:

1. If the exponent of ten is positive, move the decimal point to the right the number of digits until 10 is raised to the zero power. For example:

    $$3 \times 10^2 = 3. \times 10^2 = 30. \times 10^1 = 300 \times 10^0.$$

    Thus, $3 \times 10^2 = 300$.

    In this example, the exponent of ten was 2. Therefore, the decimal point to the right of 3 was moved two digits to the right.

2. If the exponent of ten is negative, move the decimal point to the left the number of digits until 10 is raised to the zero power. For example:

    $$4 \times 10^{-2} = 4. \times 10^{-2} = 0.4 \times 10^{-1} = 0.04 \times 10^0.$$

    Thus, $4 \times 10^{-2} = 0.04$.

    In this example, the exponent of ten was minus 2. Therefore, the decimal point to the right of 4 was moved two digits to the left.

These rules work for both positive and negative numbers.

Exercise Set 16.12

Convert the following exercises to decimal notation. (Some of these exercises are not written in scientific notation.)

1. $5 \times 10^3 =$

2. $78 \times 10^{-5} =$

3. $-392 \times 10^6 =$

4. $9267 \times 10^5 =$

5. $-7825 \times 10^7 =$

6. $-92\ 866 \times 10^{-4} =$

7. $1.932 \times 10^3 =$

8. $8.2748 \times 10^{-4} =$

9. $-3.1519 \times 10^2 =$

10. $-9.725 \times 10^{-5} =$

11. $8.42 \times 10^7 =$

12. $2.748 \times 10^{-6} =$

Your hand-held calculator has a key labeled **EE** or **EXP.** (See Appendix G.) You can use this key to enter numbers written in scientific notation directly into your calculator. Your calculator will automatically display very large and very small results using scientific notation if there is not enough room to display the decimal notation. (Read the calculator's instruction manual.)

Practice scientific notation on your hand-held calculator using the following examples.

Examples

$$3 \underset{\text{EE}}{\underline{\times 10^4}} \times 2 \underset{\text{EE}}{\underline{\times 10^{-1}}} =$$

$$4 \underset{\text{EE}}{\underline{\times 10^3}} \times 3 \underset{\text{EE}}{\underline{\times 10^2}} =$$

$$0.0002 \times 1 \underset{\text{EE}}{\underline{\times 10^0}} =$$

Numbers written using scientific notation can be multiplied and divided with or without a calculator. The rules for multiplying and dividing numbers (Chapters 4 and 5 of Book 1) and fractions (Chapter 9 of Book 1), and the laws of exponents, apply.

Study the following examples.

Example

$$(2 \times 10^2) \times (3 \times 10^3)$$
$$= 2 \times 3 \times 10^2 \times 10^3$$
$$= 6 \times 10^{2+3}$$
$$= 6 \times 10^5 \quad \text{or} \quad 600\ 000$$

Example

$$(4 \times 10^3) \times (3 \times 10^2)$$
$$= 4 \times 3 \times 10^3 \times 10^2$$
$$= 12 \times 10^{3+2}$$
$$= 12 \times 10^5$$
$$= 1.2 \times 10^6 \quad \text{or} \quad 1\ 200\ 000 \quad \text{or} \quad 1.2 \text{ million}$$

Example

$$\frac{8 \times 10^5}{4 \times 10^2}$$
$$= \frac{8}{4} \times 10^{5-2}$$
$$= 2 \times 10^3 \quad \text{or} \quad 2000$$

The rules for adding and subtracting numbers written using scientific notation are too complicated to be presented in this book. Hand-held calculators are designed to automatically apply these rules.

Exercise Set 16.13

Work the following exercises; display the results using scientific notation.

1. $2.1 \times 10^2 \times 4 \times 10^3$

2. $\dfrac{8.4 \times 10^5}{2 \times 10^3}$

3. $2.6 \times 10^5 \times 3 \times 10^{-2}$

EXPONENTS AND RADICALS

4. $\dfrac{5.7 \times 10^4}{3 \times 10^{-2}}$

5. $\dfrac{6 \times 10^5 \times 4 \times 10^2}{8 \times 10^3}$

6. $\dfrac{3.6 \times 10^5 \times 5 \times 10^2}{3 \times 10^4 \times 6 \times 10^{-3}}$

7. $\dfrac{4 \times 10^5 \times 15 \times 10^{-3}}{1 \times 10^{-4} \times 6 \times 10^7}$

Metric Unit Conversions

The business and scientific communities have agreed on a set of letter symbols called **prefixes** used to simplify the use of scientific notation and metric (SI) units. You may already be familiar with some of these symbols; they are listed in the following table along with their **multiplying factors.**

SI Prefixes and Multiplying Factors

Factor	Prefix	Symbol	Pronunciation (U.S.)	Term (U.S.)
10^{18}	exa	E	as in *Texas*	one quintillion
10^{15}	peta	P	as in *petal*	one quadrillion
10^{12}	tera	T	as in *terrace*	one trillion
10^{9}	giga	G	jig′-a (*a* as in *about*)	one billion
10^{6}	mega	M	as in *megaphone*	one million
10^{3}	kilo	k	as in *kilowatt*	one thousand
10^{-2}	centi	c	as in *sentiment*	one-hundredth
10^{-3}	milli	m	as in *military*	one-thousandth
10^{-6}	micro	μ	as in *microphone*	one-millionth
10^{-9}	nano	n	nan′-oh (as in *Nancy*)	one-billionth
10^{-12}	pico	p	peek′-oh	one-trillionth
10^{-15}	femto	f	fem′-toe (as in *feminine*)	one-quadrillionth
10^{-18}	atto	a	as in *anatomy*	one-quintillionth

Each of the sixteen **standard** symbols is known as a **prefix** because it is attached to the beginning of a unit of measure. Study the following example, where **1000 meters** is equal to **1 kilometer.**

Example   Convert 20 000 meters to kilometers.

Answer

$$20\ 000 \text{ meters}$$
$$= 20 \times 10^3 \text{ meters}$$
$$= 20 \text{ kilometers}$$

The prefix **kilo** represents 1000 or 10^3.

In our work, we will use only the SI unit prefixes that represent **10** raised to a power that is a multiple of **3.**

Study the following examples.

Example   Write the following unit of measure using an SI prefix:

$$20\ 000\ 000 \text{ watts}$$

Answer

$$= 20 \times 1\ 000\ 000 \text{ watts}$$
$$= 20 \times 10^6 \text{ watts}$$
$$= 20 \text{ megawatts}$$

*The answer could have been written as **20 000 kilowatts.***
However, the use of as few zeros as possible is the preferred result.

Example   Write the following using an SI prefix:

$$100\ 000 \text{ hours}$$

Answer

$$= 100 \times 10^3 \text{ hours}$$
$$= 100 \text{ kilohours}$$

Example   Write the following using an SI prefix:

$$0.000\ 04 \text{ second}$$

Answer

$$= 40 \times 10^{-6} \text{ second}$$
$$= 40 \text{ microseconds}$$

Note the following examples. We have rewritten the units of measure using both the SI prefix symbol and the unit symbol. (See Chapter 14 of Book 1.)

Examples

25 000 meters
= 25 km

0.03 meter
= 30 mm

0.002 second
= 2 ms

20 000 000 watts
= 20 MW

8000 meters/hour
= 8 km/h

85 000 kilobytes
= 85 megabytes

Exercise Set 16.14

Write the following units of measure using an SI prefix.

1. 0.026 gram =

2. 37 000 meters =

3. 28 200 grams =

4. 0.05 second =

5. 27 000 liters =

6. 0.04 liter =

7. 40 000 meters/hour =

8. 6000 kilobytes =

9. 5000 megabytes =

10. 186 000 miles per hour =

11. 0.000 276 milligram =

12. 0.003 548 microsecond =

Write the following units of measure using both the SI prefix symbol and the unit symbol. (See Chapter 14 of Book 1.)

13. 0.026 gram =

14. 37 000 meters =

15. 28 200 grams =

16. 0.05 second =

17. 27 000 liters =

18. 0.04 liter =

19. 40 000 meters/hour =

20. 6000 kilobytes =

21. 5000 megabytes =

22. 186 000 miles per hour =

23. 0.000 276 milligram =

24. 0.003 548 microsecond =

Radicals

If $3 \times 3 = 9$, then what are the two identical factors of **9**? The answer is 3×3, because $3 \times 3 = 9$. The answer could also be -3×-3, because $(-3)(-3) = 9$.

$$(+3)(+3) = 9$$
$$(-3)(-3) = 9$$

In 1600 B.C., a Babylonian tablet described a method for computing the two identical (positive) factors of a number. One modern math approach is to ask the question, "What is the square root of **9**?" The answer is **3**.

A symbol has been devised that indicates the square root math operation. The symbol $\sqrt{}$ is spoken **"the square root of"**; this symbol is known as a **radical** (from a Latin word meaning "root").

The radical symbol was first used in a book written in A.D. 1526 by Christoff Rudolff. The notation is as follows:

$$\sqrt[2]{9} \text{ is 3} \quad \text{or} \quad \sqrt{9} \text{ is 3}$$

Both of the above examples are spoken, "The square root of nine is three." The **2** is known as the **index** of the radical. This value **(2)** indicates the number of identical (positive) factors to be determined. Therefore, it indicates that we are to determine the square root of the number **9**. Whenever the index is omitted, a **2** is assumed to be there.

Study the following examples.

Examples

$$\sqrt{25} = \sqrt{5 \times 5} = 5$$
$$\sqrt{49} = \sqrt{7 \times 7} = 7$$

What are the three identical factors of **125**? They are $5 \times 5 \times 5$. This information helps us to answer another question: "What is the cube root of **125**?" The answer (**root** or result) is **5**:

$$\sqrt[3]{125} = \sqrt[3]{5 \times 5 \times 5} = 5$$

The first term is spoken "the cube root of **125**." The index of the radical is now **3**. The number under the radical **(125)** is known as the **radicand.** Thus,

3 is the index,

$\sqrt{}$ is the radical,

125 is the radicand, and

5 is the root (or result).

Example

$$\sqrt[3]{343} = \sqrt[3]{7 \times 7 \times 7} = 7$$

A number "raised to the second power" is said to be **squared.** For example, 3^2 is spoken "three to the second power" or "three squared." Multiplying a number by itself is known as **squaring** that number. The square of **3** is **9.** The squaring operation is indicated using exponents. For example:

$$3 \times 3 = 3^2 = 9$$

Determining the **square root** of a number is the inverse operation of squaring a number.

$$\sqrt{9} = \sqrt{3 \times 3} = 3$$

A number "raised to the third power" is said to be **cubed.** For example, 5^3 is spoken "five to the third power" or "five cubed." Multiplying three identical factors together is known as **cubing** that number. The cube of **5** is **125.** The cubing operation is indicated using exponents:

$$5 \times 5 \times 5 = 5^3 = 125$$

Determining the **cube root** of a number is the inverse of **cubing** a number:

$$\sqrt[3]{125} = \sqrt[3]{5 \times 5 \times 5} = 5$$

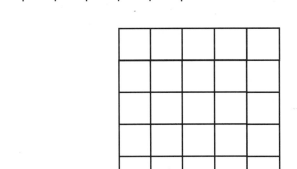

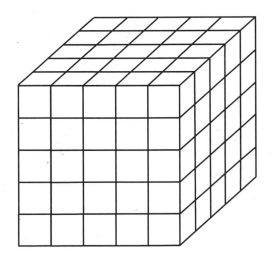

Exercise Set 16.15

Work the following exercises.

1. Determine the square root of 25.

2. What is the square root of 16?

3. What numbers when squared will result in 64?

   Note: There are two answers.

4. Determine the cube root of 27.

5. What is the cube root of 216?

6. What is the cube root of 512?

7. What is the square of 7?

8. What is the cube of 7?

9. What is the square of -2?

10. What is the cube of -2?

11. What is the value of $-(2)^2$?

MATH AT WORK

Compare the following examples:

$$\sqrt[2]{9} \qquad\qquad (9)^{1/2}$$
$$= \sqrt[2]{3 \times 3} \qquad = (3 \times 3)^{1/2}$$
$$= \sqrt[2]{3^2} \qquad = (3^2)^{1/2}$$
$$= 3 \qquad\qquad = 3^{2 \times (1/2)}$$
$$\qquad\qquad\qquad = 3^1 \quad \text{or} \quad 3$$

In the left presentation, we determined the square root of nine using the radical symbol with an index of **2.** In the right presentation, we raised nine to the one-half power, which is a fractional exponent. We then applied the fourth law of exponents (page 16–18). The results are the same.

Therefore, determining the root of a number is the same as raising that number to a **fractional exponent.** The index of the root is the reciprocal of the exponent. This fractional exponent obeys the laws of exponents, as the following example illustrates.

Example

$$\sqrt[3]{343}$$
$$= \sqrt[3]{7 \times 7 \times 7}$$
$$= \sqrt[3]{7^3}$$
$$= (7^3)^{1/3}$$
$$= 7^1 \quad \text{or} \quad 7$$

Determining the cube root of **343** is the same as raising **343** to the one-third power.

Study the following examples.

Example   What is the square root of 121?
Answer

$$\sqrt{121} = (121)^{1/2} = (11^2)^{1/2} = 11^1 = 11$$

Example   Determine the fourth root of 81.
Answer

$$\sqrt[4]{81} = (81)^{1/4} = (3^4)^{1/4} = 3^1 = 3$$

Example   What is the fifth root of 1024?
Answer

$$\sqrt[5]{1024} = (1024)^{1/5} = (2^{10})^{1/5} = 2^2 = 4$$

Note that we determined the prime factors of 1024 first before continuing the solution.

Example   Determine the cube root of 2744.
Answer

$$\sqrt[3]{2744} = (2744)^{1/3} = (2^3 \times 7^3)^{1/3} = 2 \times 7 = 14$$

EXPONENTS AND RADICALS

Note that in each example, we determined the prime factors of the radicand, then applied the laws of exponents, and then determined the root or result.

Example What is the fifth power of 2?

Answer

$$2^5 = 32$$

You may use a calculator to find any root:

Step 1:	Enter the number (radicand).
Step 2:	Depress the y^x key.
Step 3:	Enter the index.
Step 4:	Depress the **1/x** key.
Step 5:	Depress the **=** key.

Exercise Set 16.16

Work the following examples using prime factors.

1. What is the square root of 6561?

2. Determine the cube root of 3375.

3. What is the fourth root of 256?

4. Determine the square root of 169.

5. What is the fifth root of 243?

6. What are the two numbers whose square is 625?

More about Exponents and Radicals

You should be aware of the following information. You do not, at this point in your development of math knowledge, need to memorize this information.

There are three types of square roots:

1. **Exact square roots,** such as

$$\sqrt{1}, \quad \sqrt{4}, \quad \sqrt{9}, \quad \sqrt{25}$$

 The results are positive integers.

2. **Irrational square roots,** such as

$$\sqrt{2}, \quad \sqrt{3}, \quad \sqrt{5}, \quad \sqrt{6}, \quad \sqrt{7}$$

 The value of the square root is not the ratio of two integers. Therefore, these roots, when expressed as decimal fractions, are not exact or repeating numbers. They are nonrepeating decimal fractions and are known as **irrational roots.**

3. **Imaginary roots,** such as

$$\sqrt{-1}, \quad \sqrt{-2}, \quad \sqrt{-3}, \quad \sqrt{-4}, \quad \sqrt{-5}$$

 No real number when multiplied by itself can result in a negative number. Therefore, the square root of a negative number is not a real number; it is known as an **imaginary** number.

There is only one real number answer for the cube root of a number. It can be either an integer or an irrational number.

- The cube root of a positive number is a positive number.
- The cube root of a negative number is a negative number.

There can be an exponent (or power) for a number that is neither an integer nor a fraction:

- Exponents can contain decimal fractions.
- Exponents can contain irrational numbers.
- Exponents can contain imaginary numbers.
- Exponents can contain a mixture of integers, fractions, decimal fractions, irrational numbers, and imaginary numbers.

The meaning of radical exponents is discussed in some algebra books and in most science and engineering books.

Mathematicians once believed that all the rational numbers that could be imagined would **fill** the spaces between all the integers. However, Richard Dedekind in 1872 proved this theory to be wrong. It is now believed that the irrational numbers **fill** the spaces on the real number line between all the rational numbers.

EXPONENTS AND RADICALS

There is one important habit that you should acquire when you begin to use a hand-held calculator for computations involving the **EE** (or the **EXP**) key. If the first factor of a number written using scientific notation is exactly **1**, such as 1×10^4, then:

> Never write 10^4 because you may accidentally key **10 EE 4**. (Note that **10 EE 4** equals 1×10^5.)

> Always write 1×10^4. Then, even under pressure, you will key **1 EE 4**.

Hand-held calculators require the entry of the **1** (or other number) before you key **EE** (or **EXP**), followed by the exponent value.

Summary

Exponential notation is used to write the multiplication of the same numbers more efficiently. For example:

$$3 \times 3 \times 3 \times 3 = 3^4$$

where **3** is the **base** and **4** is the **exponent** or **power.**

There are five laws that govern exponential operations.

Law 1   When expressions that have the same base are multiplied together, then their exponents are added:

$$2^3 \times 2^4 = 2^7$$

Law 2   When expressions that have the same base are divided, then the exponent of the denominator is subtracted from the exponent of the numerator:

$$\frac{2^7}{2^4} = 2^{7-4} = 2^3 = 8$$

Thus, any number raised to the zero power is one:

$$\frac{5^3}{5^3} = 5^{3-3} = 5^0 = 1$$

Also, negative exponents are possible:

$$\frac{5^5}{5^7} = 5^{5-7} = 5^{-2} = \frac{1}{5^2} = 0.04$$

Law 3   When determining the reciprocal of any base that has an exponent, change the sign of that exponent:

$$3^5 = \frac{1}{3^{-5}} \quad \text{and} \quad \frac{1}{2} = 2^{-1} = 0.5$$

Law 4   When an expression consists of a base and an exponent that is raised to another exponent (power), then the exponents are multiplied together:

$$(2^3)^3 = 2^9$$

Law 5   When the product or quotient of two or more terms is raised to a power, then the exponents are multiplied by that power:

$$(2^2 \times 3^3)^3 = 2^6 \times 3^9$$

These laws apply to the multiplication or division of single-term expressions. The first two laws apply if the bases are the same. None of these laws apply to the addition or subtraction of terms.

Scientific notation combines number placement, exponential notation, and decimal numbers. Scientific notation is often referred to as **powers of ten.** The power of ten indicates how many factors of ten exist in a product (positive exponent) or quotient (negative exponent).

Examples

$$2000 = 2 \times 10 \times 10 \times 10 = 2 \times 10^3 \quad \text{(product)}$$

$$0.04 = \frac{4}{100} = \frac{4}{10 \times 10} = \frac{4}{10^2} = 4 \times 10^{-2} \quad \text{(quotient)}$$

There are international standards governing **unit prefixes.** Refer to the table on page 16–45 for the most commonly used prefixes.

One additional unit prefix is used for length measurement, such as with clothing:

$$10^{-2} \quad \text{centi} \quad \text{cm}$$

The inverse of raising a number to an integer exponent is determining the **root** or **radical** of the number. The square root ($\sqrt{\ }$) and the cube root ($\sqrt[3]{\ }$) are the most commonly used roots. Determining the root of a number is the same as raising that number to a fractional exponent. For example:

$$\sqrt[2]{9} = (9)^{1/2} = (3^2)^{1/2} = 3^1 = 3$$

The root of a number that is not exact (such as the square root of **2**) is known as an **irrational number** because no ratio of two integers exists that is exactly that number.

Glossary of Math Words

Base (bayce′)   The quantity of numbers in a number system; the base of the decimal system is **10.** In the expression 3^5, the **3** is the base.

Cube (kyoob′) **root**   The special name for the third root of a number; $\sqrt[3]{8}$ is 2.

Exponent (eks′-po-nent)   The quantity that indicates how many identical numbers are multiplied together. In the expression 2^3, the **3** is the exponent; it is also the **power** of that number, where

$$2^3 = 2 \times 2 \times 2 = 8$$

Exponential notation (eks′-po-nen′-shul no-tay′-shun)   An expression that consists of a base and an exponent. For the expression 2^4, the **2** is the base, and the **4** is the exponent (or power) of that base.

Index (in′-deks) **of a radical**   The number that indicates the root to be taken; for $\sqrt[4]{81}$, the **4** is the index.

Irrational (i-rash′-shun-ul) **number**   A number that cannot be expressed as the ratio of two integer numbers; $\sqrt{2}$ is an irrational number.

Multiplying factor (mul'-ti-ply-ing fak'-ter) The number represented by a letter prefix. *See* Prefix.

Power (pow'-ur) **of a number** *See* Exponent.

Powers of ten *See* Scientific notation.

Prefix (pree'-fiks) A letter, or group of letters, used to represent the number of 10's to be multiplied together. The measure **2 kW** or **2 kilowatts** is the same as **2 × 10³ watts**, or **2000 watts.**

Radical (rad'-i-kul) The math symbol $\sqrt{}$ indicating that a **root** of a number is to be determined.

Radicand (rad'-i-kand) The number under the **radical** symbol; **81** is the radicand of $\sqrt{81}$.

Root (root') The result of the math operation that is the inverse of raising a number to a power. The fourth root of **81** is **3**; that is, $\sqrt[4]{81} = 3$.

Scientific notation (sy-en-tif'-ik no-tay'-shun) The product of numbers from **1 to 10,** and **10** raised to a power; **3.1 × 10²** is the scientific notation for **310.**

Square (skwair) **root** The special name for the second root of a number; $\sqrt[2]{25} = \sqrt{25} = 5$.

Superscript (soo'-per-skript) A number placed above and to the right of a base. For **2³**, the number **3** is the superscript; it is also known as the **exponent.**

Chapter 16 Test

Follow instructions carefully:

> *On a separate piece of paper, write the answers to the following questions. Do* not *write on these pages.*

> *When you are finished, compare your answers with those given in Appendix B.*

> *Record the date, your test time, and your score on the chart at the end of this test.*

Simplify the following expressions using exponential notation.

1. $2 \times 2 \times 2 \times 2 \times 2 =$

2. $3 \times 3 \times 5 \times 5 \times 5 \times 5 =$

3. $\dfrac{2 \times 2 \times 2}{3 \times 3 \times 3 \times 3} =$

Calculate the values of the following expressions. Apply the laws of exponents where appropriate.

4. $2^4 \times 3^2 =$

5. $\dfrac{5^7}{5^4} =$

6. $\dfrac{4^3 \times 3^2}{2^3} =$

7. $(768)^0 =$

8. $3^2 + 2^3 =$

9. $(2^3)^2 =$

10. $(2^{-3})^2 =$

11. $(2^2 \div 3^3)^3 =$

Write the following numbers using scientific notation.

12. $50\ 000 =$

13. $0.002 =$

Determine the actual numerical values for the following.

14. $6 \times 10^2 =$

15. $4 \times 10^{-1} =$

Evaluate the following.

16. $\sqrt{36} =$

17. $\sqrt[3]{8} =$

18. $(64)^{1/3} =$

What two numbers, when squared, have 25 as their result?

19.

20.

Chapter 16 Test Record

DATE	TIME	SCORE

Refer to

- Appendix B for the correct answers to this test.
- Appendix C if taking this test required too much effort.
- Appendix D for **Memory Methods** assistance.
- Appendix E if your test scores are decreasing.

When a problem seems difficult, find one like it in this chapter. Then study that (and the related material) again.

Develop additional 3 × 5 cards for those ideas, problems, and procedures that caused you difficulty.

The Three Dimensions

One Dimension

In Chapter 14 of Book 1, we explored measure and examined units of measure, such as the inch, foot, meter, and kilometer. In this chapter we explore **dimension,** a measure that indicates length, breadth, or depth.

Units of measure can be used to describe **distance.**

Distance can be displayed along the real number line. For example, we may choose **meters** (**m**) to be the unit of measure. As shown in the following figure, starting our reference at the origin:

> Positive direction is shown to the right (east, **E**).
>
> Negative direction is shown to the left (west, **W**).

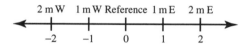

This distance is a **one-dimensional** distance; it is horizontal.

The vertical real number line can be used to describe a second direction, or dimension. As shown in the following figure, placing our reference at the origin:

> Positive direction is shown up (north, **N**).
>
> Negative direction is shown down (south, **S**).

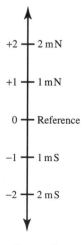

This second distance is also a one-dimensional distance; it is vertical.

Two Dimensions

When the horizontal and vertical distances are combined, then we can describe two dimensions. Navigators label these two sets of distances north, south, east, and west.

Designers often refer to these two dimensions as

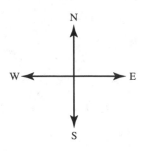

- horizontal and vertical, or
- length and width, or
- length and height, or
- length and breadth.

Three Dimensions

The third dimension, according to spaceship navigators and others, is generally considered as

- in toward our Earth's center, or
- out from our Earth's center.

Designers often refer to this third dimension as **depth.** This third distance is also a one-dimensional distance.

We can construct displays that describe the three dimensions together on a **flat, two-dimensional surface,** such as this sheet of paper. As shown in the following figure, when we want to indicate **in** or **behind** that surface, we use a dashed (-------) line:

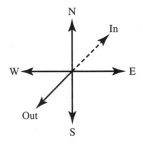

When we want to indicate **out** or **in front of** that surface, we use a solid (————————) line:

Designers often refer to these three dimensions as

- horizontal, vertical, and in-and-out; or
- length, height, and thickness; or
- length, breadth, and depth.

MATH AT WORK

Straight Lines and Circles

During the late 1700s, Jean Baptiste Fourier studied the lines or curves that were being used in the design of primitive mechanical machines. His math analysis indicated that all curves consist of

- straight lines,
- circles, or
- a combination of straight lines and circles.

The following definitions now become important to us.

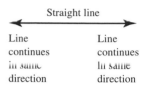

Straight Line   A **straight line** consists of an infinite number of points that, when placed side by side, go on forever in the same direction. A portion of a straight line is known as a line **segment.**

Circle   A **circle** consists of an infinite number of points that are **equidistant** (the same distance) from one other point known as the **center** of the circle. The distance from the center to any point on that circle is the **radius.** A straight line segment from a point on a circle through its center to another point on that circle is the **diameter.**

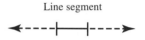

Curve   A **curve** consists of an infinite number of points that, when placed side by side, may point in different directions.

Therefore, it is more accurate to say that curves consist of

- straight lines or line segments;
- circles, or portions of circles; or
- a combination of line segments and portions of circles.

Open and Closed Surfaces

For now, we will restrict our study of curves to those that can be constructed on a flat surface. We learned in Chapter 11 of Book 1 that distance is usually measured either along a horizontal line (east and west) or along a vertical line (north and south). These two lines, shown on a flat surface, extend forever.

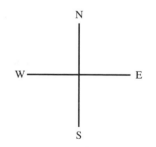

We can construct a curve on a (flat) surface. If we start at one point and end at a different point, then all of the surface near that curve is known as its **open surface.** The following figure illustrates an open surface:

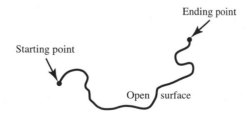

The definition of an **open surface** is not precise. Why? Because no two people have been able to agree upon the definition of the word **near. Near** is a qualitative word.

We can start constructing a curve at one point and eventually end at that same point. Then all of the surface within that curve is known as a **closed surface:**

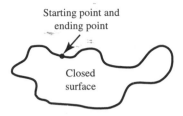

Starting point and ending point

Closed surface

There are two important measures involved with closed surfaces:

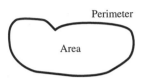

Perimeter

Area

1. The distance along the curve from its starting point to its ending point is known as the surface **perimeter.**
2. The surface enclosed by a perimeter is known as the surface **area.**

Perimeter and Surface Area of Common Shapes

We will next examine the perimeters and areas of only the most common and simple closed surfaces. As we examine each shape, you may find it helpful to identify an item around you with that shape. Ask a friend to help, and see how many items you can identify.

Square   The simplest closed (or **enclosed**) surface is a square. A square has four sides of equal length; these sides are either horizontal or vertical. A checkerboard contains examples of squares.

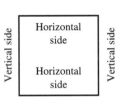

Vertical side | Horizontal side / Horizontal side | Vertical side

The perimeter (*P*) of a square equals the sum of the length of its four sides. The perimeter (*P*) of the following square equals the sum of its four sides:

$$\text{Perimeter} = \text{side} + \text{side} + \text{side} + \text{side}$$
$$P = 2 \text{ cm} + 2 \text{ cm} + 2 \text{ cm} + 2 \text{ cm}$$
$$= 8 \text{ cm}$$

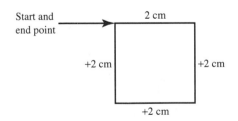

Start and end point

2 cm

+2 cm   +2 cm

+2 cm

Recall from Chapter 14 of Book 1 that **cm** is the abbreviation of the metric measure **centimeter.**

Most persons soon realize that you can determine the perimeter of a square more quickly by multiplying the length of one side (2 cm) by four, as shown in the following figure:

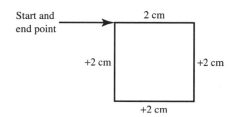

Thus,

$$P = 4 \times 2 \text{ cm}$$
$$= 8 \text{ cm}$$

Be certain to include the units of measure in your computations.

The area of a square is the most fundamental measure of a closed surface. All units of area include the word **square.** The area (*A*) of a square is the product of any two sides:

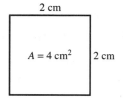

$$\text{Area} = (\text{side}) \times (\text{side})$$
$$A = (2 \text{ cm}) \times (2 \text{ cm})$$
$$= 4 \text{ cm}^2$$

The result is spoken "**four square centimeters.**" It can be written as either 4 cm² or 4 sq cm.

$$A = 2 \text{ cm} \times 2 \text{ cm}$$

$$A = 4 \text{ sq cm} \quad \text{or} \quad 4 \text{ cm}^2$$

There are three measures related to a square:

1. the length of each side,
2. the perimeter, and
3. the area.

When any one of these three measures is known, then it is possible to determine the other two missing measures.

Study the following examples.

Example   The side of a square measures 5 cm. Determine its perimeter (P) and its area (A).

Answer

$$P = 4 \times (5 \text{ cm})$$
$$= 20 \text{ cm}$$

$$A = (5 \text{ cm}) \times (5 \text{ cm})$$
$$= 25 \text{ cm}^2$$

Example   The perimeter of a square is 64 cm. Determine the length of a side (s) and its area (A).

Answer

$$s = \frac{64 \text{ cm}}{4}$$
$$= 16 \text{ cm}$$

$$A = (16 \text{ cm}) \times (16 \text{ cm})$$
$$= 256 \text{ cm}^2$$

Example   The area of a square is 36 cm^2. Determine the length of a side (s) and its perimeter (P).

Answer

$$s = \sqrt{36 \text{ cm}^2}$$
$$= 6 \text{ cm}$$

$$P = 4 \times (6 \text{ cm})$$
$$= 24 \text{ cm}$$

You have two options regarding the procedures illustrated in these examples:

1. Memorize the procedure for solving each of these problems.
2. Examine each type of problem and determine the reasoning (logic) that led to its solution.

As you learn to reason, you will then acquire a deeper understanding of the math concepts involved.

Exercise Set 17.1

The answers to the exercises in this chapter are given in Appendix A.

Work the following exercises; sketch each square.

1. A square floor tile is 7 cm on one side. Determine its perimeter and its area.

2. A square is 12 inches on one side. Determine its perimeter and its area.

3. A square has a side that is 8 cm long. Determine its perimeter and its area.

4. The perimeter of a square is 28 inches. Determine the length of one side and its area.

5. The area of a square is 81 cm^2. Determine the length of one side and its perimeter.

6. The perimeter of a square is 14 inches. Determine the length of one side and its area.

MATH AT WORK

Rectangle A **rectangle** is another simple closed surface. It consists of two horizontal sides, usually known as the lengths (ℓ), and two vertical sides, usually known as the widths (w). Its perimeter (P) is the sum of the four sides:

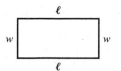

$$P = \ell + w + \ell + w$$

You can determine the perimeter more quickly by adding twice the length to twice the width:

$$P = 2 \times \ell + 2 \times w$$

Examine the rectangle shown in the following figure:

$\ell = 4$ in.

$w = 2$ in. $w = 2$ in.

$\ell = 4$ in.

Thus,

$$P = 2 \times 4 \text{ in.} + 2 \times 2 \text{ in.}$$
$$= 12 \text{ in.}$$

Several centuries ago, a mathematician proved that the area (A) of a rectangle is the product of its length (ℓ) and its width (w):

$$A = \ell \times w$$
$$= (4 \text{ in.}) \times (2 \text{ in.})$$
$$= 8 \text{ sq in.} \quad \text{or} \quad 8 \text{ in.}^2$$

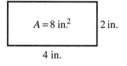

We can show that there are two squares within this simple rectangle. Each square has an area of 2 in. $\times$ 2 in. or 4 in.^{2}. The sum of the area of these two squares and, therefore, the rectangle is

$$A = 4 \text{ in.}^2 + 4 \text{ in.}^2$$
$$= 8 \text{ in.}^2$$

THE THREE DIMENSIONS

We are now developing **formulas.** We are using letters to describe the relation between two or more quantities. Thus, for sides (s), lengths (ℓ), widths (w), perimeters (P), and areas (A), we now have four formulas that relate to squares and rectangles. These formulas are listed in the following table:

	Square	**Rectangle**
Perimeter (P)	$P = 4 \times s$	$P = 2 \times \ell + 2 \times w$
Area (A)	$A = s^2$	$A = \ell \times w$

Note the similarities between the formulas for the perimeter and area of the rectangle and the square.

We will provide you with these formulas when you need them. They are also shown in the following figure:

Exercise Set 17.2

Work the following exercises; sketch each rectangle.

1. A rectangle has a length of 14 cm and a width of 6 cm. Determine its perimeter and its area.

2. The two sides of a rectangle are 5 in. and 3 in. Determine its perimeter and its area.

3. The area of a rectangle is 40 cm^2; one side is 5 cm long. Determine the perimeter of the rectangle and its other dimension.

4. The sides of a rectangle are measured to be 5.3 cm and 2.7 cm. Determine its perimeter and its area.

Triangle A **triangle** is a closed surface that has three sides. If one side of a triangle is forced to be horizontal, and if the vertical distance from its highest point (peak) to the horizontal side (**base**) is known, then the computations are simple.

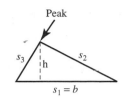

Label the triangle dimensions as follows:

- The sides are s_1, s_2, and s_3.
- The **base** (horizontal side) is s_1 or b.
- The **altitude** or **height** (vertical distance from the peak to the base) is h.

Then the formulas for perimeter (P) and area (A) are as follows:

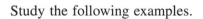

$$\text{Perimeter} = \text{first side} + \text{second side} + \text{third side}$$
$$P = s_1 + s_2 + s_3$$

$$\text{Area} = \frac{1}{2}(\text{base}) \times (\text{height})$$

$$A = \frac{1}{2}bh$$

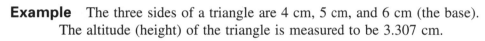

Study the following examples.

Example The three sides of a triangle are 4 cm, 5 cm, and 6 cm (the base). The altitude (height) of the triangle is measured to be 3.307 cm.

    a. Determine its area and its perimeter.

    b. Sketch and label the triangle.

Answer

    a. $P = 4 \text{ cm} + 5 \text{ cm} + 6 \text{ cm}$
        $= 15 \text{ cm}$

        $A = \frac{1}{2} \times (6 \text{ cm}) \times (3.307 \text{ cm})$

        $= 9.921 \text{ cm}^2$

    b.

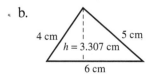

Example   A triangle has a base of 3 in., a second side (that is also its altitude) of 4 in., and a third side of 5 in. This is a special type of triangle, known as a **right triangle,** because one of the sides is also the triangle altitude.

    a. Determine its perimeter and area.

    b. Sketch and label this triangle.

Answer

    a. $P = 3 \text{ in.} + 4 \text{ in.} + 5 \text{ in.}$

          $= 12 \text{ in.}$

        $A = \dfrac{1}{2} \times (3 \text{ in.}) \times (4 \text{ in.})$

          $= 6 \text{ in.}^2$

    b.

Exercise Set 17.3

Work the following exercises; sketch and label the triangles.

1. The three sides of a triangle are 9 cm (the base), 7 cm, and 4 cm. The altitude is measured; it is 2.981 cm. Determine the perimeter and the area of the triangle.

2. The sides of a triangle are 10 in., 10 in., and 16 in. (the base). The altitude is 6 in. Determine the perimeter and the area of the triangle.

3. A triangle's sides are 4.72 cm, 8.61 cm (the base), and 6.43 cm; the altitude is 3.47 cm. Determine the perimeter and the area of this triangle.

4. A triangle's altitude measures to be 4.822 in.; its base is 17 in., and its other two sides are 9 in. and 22 in. Determine the triangle's perimeter and area.

Parallelogram A **parallelogram** is a four-sided closed surface whose opposite sides have the same length and are in the exact same direction. These opposite sides are known as **parallel** sides. The formula for the perimeter of a parallelogram is identical to that for the rectangle. The perimeter (P) is the sum of both pairs of parallel sides:

$$P = 2 \times \ell + 2 \times w$$

The formula for the area of a parallelogram is similar to the formula for the area of a rectangle. Why? Because you can cut off the triangle-shaped portion of one end of a parallelogram and fit it exactly onto the other end. Then you have a rectangle. It can then be shown that the area of this rectangle is identical to the area of the original parallelogram. The following figures illustrate this concept:

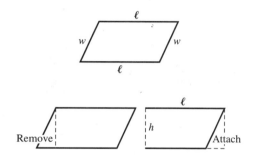

Therefore, the formula for the area (A) of a parallelogram is

$$A = \ell \times h$$

where h is the **height** of the parallelogram; it is similar to the height or altitude of a triangle.

Study the following example.

Example The sides of a parallelogram are 5 in. and 8 in. (horizontal side, or base). Determine the perimeter and the area if the height is 3 in.

Answer

$$P = 2 \times 8 \text{ in.} + 2 \times 5 \text{ in.}$$
$$= 16 \text{ in.} + 10 \text{ in.}$$
$$= 26 \text{ in.}$$

$$A = 8 \text{ in.} \times 3 \text{ in.}$$
$$= 24 \text{ in.}^2$$

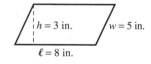

Exercise Set 17.4

Work the following exercises; sketch the parallelogram.

1. The sides of a parallelogram are 18 in. and 24 in. (base). Determine the perimeter and the area if the height is 15 in.

2. The sides of a parallelogram are 4.6 cm and 7.8 cm (base). Determine its perimeter and its area if the height is 3.2 cm.

3. The sides of a parallelogram are 3.4 ft and 6.7 ft (base). Determine its perimeter and its area if the height is 2.6 ft.

4. The sides of a parallelogram are 5.7 mm and 9.3 mm (base). Determine its perimeter and its area if the height is 4.4 mm.

Circle   As early as 2000 B.C., the measures involving a circle fascinated and confused geometers (mathematicians). We indicated at the beginning of this chapter that a circle's definition requires the existence of a center. The following figure illustrates the parts of a circle:

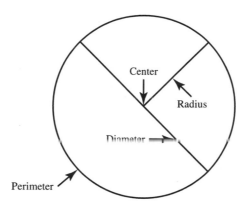

From the center of a circle, a **radius** (*r*) may be constructed.

Through the center of a circle, a **diameter** (*d*) may be constructed. Note that the diameter is twice the length of the radius; the radius is, therefore, one-half the length of the diameter.

The **perimeter** of a circle appears so frequently in computations that it has been given a special name: **circumference;** the letter symbol used is *C*.

$$\text{Perimeter} = \text{circumference} = C$$

Many attempts were made to determine the exact circumference of a circle once the radius (or diameter) was known. Crude measurements in 2000 B.C. resulted in a circumference-to-diameter ratio (*C/d*) of **3/1** or **3.** In 250 B.C., Archimedes discovered that the circumference-to-diameter ratio was not the ratio of two integers. Therefore, *C/d* is an irrational number. This irrational number was assigned the Greek letter symbol π (spoken "pie") because it is a nonrepeating decimal fraction. Our ability to determine the value of π has greatly improved. Its value has been determined by modern computers to hundreds of decimal places. However, it is not practical to apply all of these digits. Thus, we will use

$$\pi = 3.14159$$

Your calculator will automatically apply its stored value of π to any computation where you use the π key.

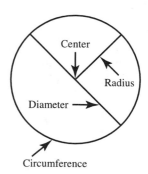

The formula for the perimeter (circumference) of a circle is

$$\text{Circumference} = (\pi) \times (d)$$
$$C = \pi d$$

When we compute the area of a circle, π is again involved. This time, the ancient mathematicians found that the value of π and the circle radius were all that had to be known.

$$\text{Area} = (\pi) \times (\text{radius})^2$$
$$A = \pi r^2$$

Study the following example.

Example The radius of a circle is 6 cm. Determine its circumference and its area.

Answer

$$C = \pi d \quad \text{where } d = 2r$$
$$= 2 \times 6 \text{ cm}$$
$$= 12 \text{ cm}$$

Therefore,

$$C = \pi \times (12 \text{ cm})$$
$$= 37.699 \text{ cm}$$

and

$$A = \pi r^2$$
$$= \pi \times (6 \text{ cm})^2$$
$$= 36\pi \text{ cm}^2$$
$$= 113.097 \text{ cm}^2$$

The formulas for circumference and area of a circle can be revised so that either the radius or the diameter can be used.

$$C = \pi d \quad \text{where } d = 2r$$
$$= \pi \times (2r)$$
$$= 2\pi r$$

and

$$A = \pi r^2 \quad \text{where } r = d/2$$

Thus,

$$A = \pi \left(\frac{d}{2}\right)^2$$
$$= \frac{\pi d^2}{4}$$

Again, it is not necessary to memorize these two different versions of the circumference and area formulas; they will be available to you when you need them.

Exercise Set 17.5

Work the following exercises; sketch and label the circles.

1. Determine the circumference and the area of a circle whose radius is 7 cm.

2. Determine the circumference and the area of a wheel whose diameter is 18 in.

3. Determine the circumference and the area of a tire whose diameter is 19.72 cm.

4. Determine the radius and the area of a wreath whose circumference is 34.8 cm.

5. Determine the diameter and the area of a circle whose circumference is 81.3 in.

Perimeter, Surface Area, and Volume

Our observable world consists of three dimensions. These three dimensions are

> horizontal,
>
> vertical, and
>
> either toward or away from an observer.

A **cube,** such as a child's block, is the simplest object that is described using these three dimensions.

These three dimensions are normally noted as

> length, or base;
>
> width, or breadth; and
>
> height, or depth.

A three-dimensional cube has twelve edges or sides (s) and six surfaces. The length, the width, and the height of a cube are all equal. Therefore, all twelve sides have the same measure, and all six surfaces are squares.

A three-dimensional object, such as a cube, seldom has a perimeter that is of interest. The twelve identical side measurements could be combined into a **perimeter.** This value would be useful if you wanted to tape these sides or fill the junction (connection) between the surfaces.

Note that a cube has six enclosed and attached surfaces. Each of these surfaces has its own area. These six surface areas of a cube are equal in value. Therefore, the total surface area (A) of a cube is

$$A = 6s^2$$

The total surface area is of interest if someone wants to paint, or place a covering around, a cube.

The six surfaces of a cube enclose a **volume** (V) of space. A cube is the simplest possible volume that can be enclosed. If the unit of side (edge) measure is the centimeter, then the following measures apply:

- Length is measured in centimeters (cm).
- Area is measured in square centimeters (cm^2).
- Volume is measured in cubic centimeters (cm^3).

The formula for the volume (V) of a cube is

$$V = s^3$$

THE THREE DIMENSIONS

Study the following example.

Example A cube measures 3 cm on a side. Compute the total surface area
(A) and the enclosed volume (V).

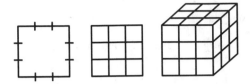

Answer

$$A = 6s^2$$
$$= 6 \times 9 \text{ cm}^2$$
$$= 54 \text{ cm}^2$$

$$V = s^3$$
$$= (3 \text{ cm})^3$$
$$= 27 \text{ cm}^3$$

*Note: 27 cm³ is spoken "**twenty-seven cubic centimeters.**" It
can also be written as 27 cu cm.*

Exercise Set 17.6

Work the following exercises; sketch and label each cube.

1. A cube measures 5 in. on a side. Compute its (total) surface area and its volume.

2. A cube measures 7 cm on a side. Compute its surface area and its volume.

3. The surface area of a cube is 64 cm^2. Compute its side dimensions and its volume.

4. The volume of a cube is 216 cm^3. Compute its side dimensions and its surface area.

5. The surface area of a cube is 625 square inches. Compute its side dimensions and its volume.

6. The volume of a cube is 343 cubic inches. Compute its side dimensions and its surface area.

7. One side of a cube measures 8 in. Compute its surface area and its volume.

A cube-shaped enclosure that has rectangular surfaces is known as a **rectangular parallelepiped.** Each surface area (A_1, A_2, A_3) of a side of a parallelepiped is determined using the rectangle formula, adapted to the labeling of the sides of the parallelepiped, as shown in the following figure:

$$A_1 = \ell \times w$$

$$A_2 = w \times d$$

$$A_3 = d \times \ell$$

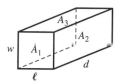

The **opposite** surfaces (A_4, A_5, A_6) have the same surface areas as A_1, A_2, and A_3. Therefore, the surface area (A) of a rectangular parallelepiped is the sum of the six rectangular surfaces:

$$A = 2 \times \ell \times w + 2 \times w \times d + 2 \times d \times \ell$$
$$= 2(\ell \times w + w \times d + d \times \ell)$$

The volume (V) of a parallelepiped is

$$\text{Volume} = (\text{length}) \times (\text{width}) \times (\text{depth})$$
$$V = \ell \times w \times d$$

Note the similarity of these two formulas to those of a cube.

Study the following example.

Example Determine the individual surface areas, the total surface area, and the volume of a rectangular parallelepiped. The three sides measure 4.6 cm, 6.2 cm, and 8.3 cm, as shown in the figure on the right.

Answer

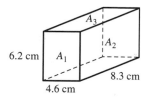

$$A_1 = (4.6 \text{ cm}) \times (6.2 \text{ cm})$$
$$= 28.52 \text{ cm}^2$$

$$A_2 = (6.2 \text{ cm}) \times (8.3 \text{ cm})$$
$$= 51.46 \text{ cm}^2$$

$$A_3 = (8.3 \text{ cm}) \times (4.6 \text{ cm})$$
$$= 38.18 \text{ cm}^2$$

$$\text{Total area} = 2 \times (28.52 + 51.46 + 38.18) \text{ cm}^2$$
$$= 236.32 \text{ cm}^2$$

$$V = (4.6 \text{ cm}) \times (6.2 \text{ cm}) \times (8.3 \text{ cm})$$
$$= 236.716 \text{ cm}^3$$

Exercise Set 17.7

Work the following exercises; sketch the parallelepiped.

1. Determine the (total) surface area and the volume of a parallelepiped whose three sides measure 6 in., 8 in., and 12 in.

2. Determine the surface area and the volume of a parallelepiped whose three sides measure 8.1 cm by 9.3 cm by 6.6 cm.

3. One **face** (surface) of a parallelepiped is a square whose area is 49 cm^2. The third side is 9 cm. Determine the measurements of the other two sides, the (total) surface area, and the volume.

4. A parallelepiped has one face whose surface area is 63 cm^2 with one of its sides measured to be 7 cm. The third side is measured to be 12 cm. Determine the missing dimension, the (total) surface area, and the volume of this parallelepiped.

MATH AT WORK

Sphere A **sphere** is the three-dimensional version of a circle; it looks like a baseball. The earth is almost spherical in shape.

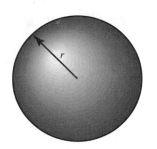

Each point on the surface of a sphere is at exactly the same distance from its center. There is no perimeter for a sphere.

Any line segment from the sphere's center to its surface is known as a **radius** (*r*). A sphere has a **surface area** (*A*) and a **volume** (*V*):

$$A = 4\pi r^2$$

$$V = \frac{4}{3}\pi r^3$$

A sphere is described as **round.** If you were to carefully cut through a sphere, and if the cut passed through the center of the sphere, then the result would be two halves of a sphere. Each half is known as a **hemisphere.** The flat surface of each hemisphere is bounded by a circle. The radius of either circle is the radius of both the sphere and the hemisphere. A hemisphere has a circumference because it has one flat, circular surface, as shown in the following figure:

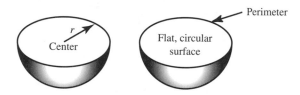

Its surface area is one-half the surface area of a sphere plus the area of the flat, circular surface:

$$A = \frac{1}{2} \cdot 4\pi r^2 + \pi r^2$$
$$= 2\pi r^2 + \pi r^2$$
$$= 3\pi r^2$$

Its volume is one-half the volume of a sphere:

$$V = \frac{1}{2} \cdot \frac{4}{3}\pi r^3 = \frac{2}{3}\pi r^3$$

Study the following examples.

Example Determine the surface area and the volume of a sphere whose radius is 6 cm.

Answer

$$A = 4\pi(6 \text{ cm})^2$$
$$= 144\pi \text{ cm}^2$$
$$= 452.4 \text{ cm}^2$$

$$V = \frac{4}{3}\pi(6 \text{ cm})^3$$
$$= 288\pi \text{ cm}^3$$
$$= 904.8 \text{ cm}^3$$

Example Determine the surface area and the volume of a hemisphere whose radius is 6 cm.

Answer

$$\text{Hemisphere } A = 3\pi r^2$$
$$= 3\pi(6 \text{ cm})^2$$
$$= 339.29 \text{ cm}^2$$

$$\text{Hemisphere } V = \frac{2}{3}\pi r^3$$
$$= \frac{2}{3}\pi(6 \text{ cm})^3$$
$$= 452.39 \text{ cm}^3$$

Exercise Set 17.8

Work the following exercises.

1. Determine the surface area and the volume of a ball whose radius is 5 cm.

2. Determine the surface area and the volume of a melon whose radius is 4 in.

3. Determine the surface area and the volume of one hemisphere whose radius is 8 cm.

4. Determine the surface area and the volume of one hemisphere whose radius is 14 in.

Rod How can you simply describe a rod? If a **rod** is placed on a horizontal surface, as shown in the following figure, then the vertical surface is either a flat circle or a curved surface that looks like a rectangle or a square. A soup can is a close example to a rod.

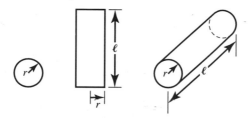

The surface of a rod consists of three parts:

1. two flat, circular surfaces whose areas combine to be $2 \times (\pi r^2)$, and
2. one curved surface whose area is $2\pi r \ell$.

Thus, the total surface area (A) of a rod is

$$A = 2\pi r^2 + 2\pi r \ell$$

The **volume** (V) of rod is computed using a much simpler formula:

$$V = \pi \ell r^2$$

Study the following example.

Example Determine the surface area and the volume of a rod whose radius is 3 cm and whose length is 7 cm.

Answer

$$
\begin{aligned}
A &= 2\pi(3 \text{ cm})^2 + 2\pi(3 \text{ cm}) \times (7 \text{ cm}) \\
&= 56.55 \text{ cm}^2 + 131.9 \text{ cm}^2 \\
&= 188.5 \text{ cm}^2
\end{aligned}
$$

$$
\begin{aligned}
V &= \pi(7 \text{ cm}) \times (3 \text{ cm})^2 \\
&= 197.92 \text{ cm}^3
\end{aligned}
$$

Exercise Set 17.9

Work the following exercises.

1. Determine the surface area and the volume of a rod whose radius is 5 cm and whose length is 9 cm.

2. Determine the surface area and the volume of a rod whose radius is 4 in. and whose length is 8 in.

3. Determine the surface area and the volume of a rod whose radius is
   12.5 cm and whose length is 17 cm.

4. Determine the surface area and the volume of a rod whose radius is
   75.6 in. and whose length is 52.11 in.

MATH AT WORK

Summary

All **curves** consist of

- **straight lines** or **line segments,**
- **circles** or portions of circles, or
- a combination of line segments and portions of circles.

Straight lines and circles are the simplest two-dimensional curves:

A **straight line** consists of an infinite number of points that when placed side by side go on forever in the same direction. (A portion of a straight line is known as a **line segment.**)

A **circle** consists of an infinite number of points that are equidistant from one other point known as the **center** of the circle.

Two-dimensional curves are usually displayed on a **flat surface.**

A curve that starts at one point and neither touches nor crosses does not enclose a portion of a surface. The surface near the curve is known as an **open surface.**

A curve that starts and ends at the same point encloses a surface. This surface is known as a **closed surface:**

The distance along a curve is known as the **perimeter** of that curve.

The surface enclosed by the perimeter of a curve is known as the **surface area.**

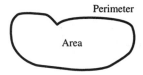

Formulas have been determined for the more common enclosed-curve shapes so that their areas and perimeters may be computed. The formulas are given in the following figure:

Shape and Name	One Dimension Name	Perimeter and Area Formulas
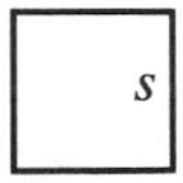 Square	s: side	$P = 4s$ $A = s^2$
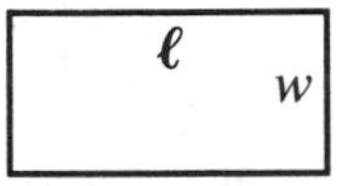 Rectangle	ℓ: length w: width	$P = 2\ell + 2w$ $A = \ell w$
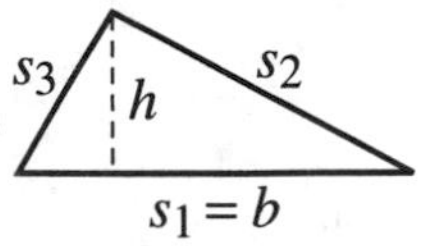 Triangle	s_1, s_2, s_3: sides b: horizontal h: vertical	$P = s_1 + s_2 + s_3$ $A = \frac{1}{2} bh$
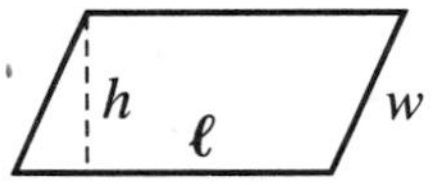 Parallelogram	ℓ: length w: width h: height	$P = 2\ell + 2w$ $A = \ell h$
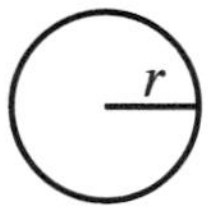 Circle	r: radius d: diameter $d = 2r$	$P = 2\pi r$ $A = \pi r^2$

The more common three-dimensional enclosed shapes sometimes have perimeters; they always have a **surface area** and an enclosed **volume.** These formulas are given in the following figure:

Shape and Name	Perimeter and Area Formulas	Volume Formula
Cube	$P = 12s$ $A = 6s^2$	s^3
Parallelepiped	$P = 4(\ell + w + d)$ $A = 2(\ell w + wd + d\ell)$	ℓwd
Sphere	(no perimeter) $A = 4\pi r^2$	$\frac{4}{3}\pi r^3$
Rod	$P = 4\pi r$ $A = 2\pi r^2 + 2\pi r\ell$	$\pi \ell r^2$

These nine (two-dimensional plus three-dimensional) shapes, or approximations to these shapes, frequently occur in our everyday world. These exact shapes, and others, are explored in greater detail in courses usually titled "Geometry."

Glossary of Math Words

Altitude (al′ ti tood)   The vertical edge of a two-dimensional surface.

Area (air′-ee-a)   A two-dimensional surface measurement; the surface enclosed by a curve's perimeter.

Base (bayce′)   The horizontal edge of a two-dimensional surface.

Center (sen′-ter) **of a circle**   The point in a circle from which all points on the circumference are equidistant.

Circle (sur′-kl)   A line represented by an infinite number of points that are equidistant from one other point known as the **center of the circle.**

Circumference (sur-kum′-fur-ens) **of a circle**   The distance from one point, around the circle, and back to the same point; the circle's **perimeter.**

Closed (clozd′) **surface**   A two-dimensional area within a curve that starts at a point and ends at that same point.

Cube (kyoob′)   A three-dimensional object with six square surface areas that enclose a volume.

Curve (kerv′)   An infinite number of points, placed side by side. A curve can consist of (1) straight lines or line segments, (2) circles or portions of circles, or (3) a combination of line segments and portions of circles.

Curved surface (kervd′ sur′-fis)   A two-dimensional area that is not flat; the surface area of a sphere is a curved surface.

Diameter (dy-am′-it-er)   A line segment that starts at one point on the perimeter of a circle, passes through the center of the circle, and ends at another point on that perimeter. A diameter divides a circle into two equal halves; the diameter is twice the length of the radius.

Dimension (di-men′-shun)   A measure that indicates length, or breadth, or depth.

Dimension, one   A single measure that usually indicates length, or breadth, or depth.

Dimensions, two   Two measures that usually indicate length and breadth.

Dimensions, three   Three measures that usually indicate length, breadth, and depth. A third dimension is indicated on a flat surface as a dashed (------) line when it is shown behind that surface.

Distance (dis′-tense)   A measure along either a line or a curve.

Equidistant (ee′-kwe-dis′-tent)   The same distance.

Flat (flat) **surface**   A two-dimensional area that is not curved; the surface area of a square is a flat surface.

Height (hyte)   The vertical edge of a two-dimensional surface.

Line (lyne) **segment**   A portion of a straight line that has a definite beginning and a definite end.

Open (oh′-pen) **surface**   A two-dimensional area; the surface near a curve that starts at one point and ends at a different point.

Parallelepiped (par-uh-lel′-uh-pie′-ped)   A three-dimensional object with six square or rectangular surfaces that enclose a volume.

Parallel lines (par′uh-lel lynes)   Lines that are in the exact same direction.

Parallelogram (par-uh-lel′-uh-gram)   A four-sided, two-dimensional object with parallel and equal-length opposite sides that enclose a surface.

Perimeter (per-im′-it-er)   The distance along a curve from a starting point back to that same point.

Radius (ray′-dee-us)   A line segment that starts at the center of a circle and ends at a point on that circle's circumference. The radius is one-half the length of the diameter.

Rectangle (rek′-tan-gl)   A four-sided, two-dimensional object whose opposite sides are equal in length, are either horizontal or vertical, and enclose a surface.

Rod (rod)   A three-dimensional object with two opposite surfaces that are circles and a third surface that is a rectangle.

Segment (seg′-ment)   A portion of a line or curve that has both a starting point and an ending point.

Sphere (sfeer′)   A three-dimensional object whose surface consists of an infinite number of points that are equidistant from its center; a baseball is nearly a sphere.

Square (skwair′)   A four-sided, two-dimensional object whose sides are equal in length, are either horizontal or vertical, and enclose a surface.

Straight (strayt) **line**   An infinite number of points that, when placed side by side, go on forever in the same direction.

Surface (sur′-fis)   A two-dimensional area that is either closed or open; it may be either flat or curved.

Triangle (try′-an-gl)   A three-sided object that is an enclosed surface.

Volume (vol′-yoom)   A three-dimensional measure that is enclosed by one or more two-dimensional surfaces.

Chapter 17 Test

Follow instructions carefully:

> *On a separate piece of paper, write the answers to the following questions. Do* not *write on these pages.*
>
> *When you are finished, compare your answers with those given in Appendix B.*
>
> *Record the date, your test time, and your score on the chart at the end of this test.*

1. The distance from a starting point on a curve, along that curve, and back to the starting point is known as the curve _____.

2. The surface enclosed by a curve perimeter is known as the curve's surface _____.

3. A square has a side that is 3.1 cm long. Its perimeter is _____.

4. A square has a side that is 3.1 cm long. Its area is _____.

5. A rectangle measures 4.2 cm on one side and 6.1 cm on the other side. Its perimeter is _____.

6. A rectangle measures 4.2 cm on one side and 6.1 cm on the other side. Its area is _____.

A triangle has a base of 3.6 in., a second side of 2.1 in., and a third side of 4.5 in. Its altitude is 2.06 in.

7. Compute the perimeter of this triangle.

8. Compute the area of this triangle.

9. The sides of a parallelogram are 4.9 cm and 8.3 cm (base). Determine the perimeter.

10. The sides of a parallelogram are 8.3 cm (base) and 4.9 cm. Determine the area if the height is 3 cm.

11. The radius of a circle is 5.6 cm. Determine the circle's circumference.

12. The radius of a circle is 9.3 cm. Determine the circle's area.

13. A cube is 4.2 in. on one side. Determine the cube's surface area.

14. A cube is 4.2 in. on one side. Determine the cube's volume.

15. A parallelepiped has the following dimensions: 4.6 cm by 3.7 cm by 9.2 cm. Determine its surface area.

16. A parallelepiped has the following dimensions: 4.6 cm by 3.7 cm by 9.2 cm. Determine its volume.

17. A sphere has a radius of 4.6 in. Determine the sphere's surface area.

18. A sphere has a radius of 4.6 in. Determine the sphere's volume.

19. A rod has a radius of 3.7 cm and a length of 9.1 cm. Determine the surface area of the rod.

20. A rod has a radius of 3.7 cm and a length of 9.1 cm. Determine the rod's volume.

Chapter 17 Test Record

DATE	TIME	SCORE

Refer to

- Appendix B for the correct answers to this test.
- Appendix C if taking this test required too much effort.
- Appendix D for **Memory Methods** assistance.
- Appendix E if your test scores are decreasing.

When a problem seems difficult, find one like it in this chapter. Then study that (and the related material) again.

Develop additional 3 × 5 cards for those ideas, problems, and procedures that caused you difficulty.

Part 5 Review Test

CHAPTER 15
More Fractions

$$\frac{\frac{2}{5} \times \frac{3}{7}}{\frac{4}{7} \div \frac{6}{5}}$$

CHAPTER 16
Exponents and Radicals

$7^2 = 49$ $\sqrt{25} = 5$

$\sqrt{25} = 5$ $7^2 = 49$

CHAPTER 17
The Three Dimensions

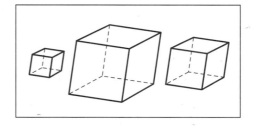

Instructions

Review your 3 × 5 cards before you take this test.

Complete this test.

You may find some ideas, problems, or procedures difficult or confusing. Develop 3 × 5 cards for them.

Part 5 Review Test

1. Simplify and reduce the following fraction:

$$\frac{8/3}{18/7}$$

2. Simplify and reduce the following fraction:

$$\frac{\frac{4}{7} \times \frac{3}{4}}{\frac{1}{2} \div \frac{7}{12}}$$

3. Simplify and reduce the following fraction:

$$\frac{\frac{2}{3} - \frac{1}{2}}{\frac{1}{2} - \frac{1}{3}}$$

4. Simplify and reduce the following fraction:

$$\frac{\frac{4}{7} \div \frac{3}{4}}{\frac{1}{2} \times \frac{7}{12}}$$

5. Simplify and reduce the following fraction:

$$\frac{\frac{2}{3} + \frac{1}{2}}{\frac{1}{2} + \frac{1}{5}}$$

6. Calculate the value represented by the following expression:

$$7^5 \div 7^3$$

7. Calculate the value represented by the following expression:

$$(3^2 \times 4)^3$$

8. Write the following number using scientific notation:

603 000

9. Determine the numeric value for the following expression:

4×10^{-2}

10. Evaluate the following expression:

$\sqrt[3]{27}$

11. Evaluate the following expression:

$(16)^{1/2}$

12. The perimeter of a circle is known as its _____.

13. The line from the center of a circle to its perimeter is known as its _____.

The perimeter of a square is 28 in.

14. Each of its sides measures _____.

The diameter of a circle is 8.32 cm.

15. Its perimeter (circumference) is _____.

16. Its area is _____.

A cube is 2.89 in. on one side.

17. Its volume is _____.

18. Its total surface area is _____.

A sphere has a radius of 2.749 cm.

19. Its volume is _____.

20. Its total surface area is _____.

Part 5 Review Test Record

DATE	TIME	SCORE

Refer to

- Appendix B for the correct answers to this test.
- Appendix C if taking this test required too much effort.
- Appendix D for **Memory Methods** assistance.
- Appendix E if your test scores are decreasing.

When a problem seems difficult, find one like it near the page referenced with the answers in Appendix B. Study the related material.

Develop additional 3 × 5 cards for those ideas, problems, and procedures that caused you difficulty.

$$\frac{4+8}{20} = 0.60$$

$$\text{or} \quad 60\%$$

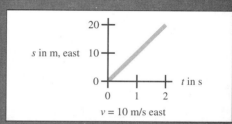

$v = 10$ m/s east

Averages and Percent

Review of Ratios and Rates

In Chapter 9 of Book 1, we examined fractions. The top of a fraction is known as the **numerator;** the bottom of a fraction is known as the **denominator.** Fractions are **ratios** or **rates.** We compare the value of the numerator to the value of the denominator.

In Chapter 14 of Book 1, we added the additional concept of the time-related rates, that is, speed, velocity, acceleration, and pay. Recall that they involve time in their denominators.

In Chapter 14, we also examined the slope and road-grade ratios. These ratios have no units.

In this chapter we will examine more ratios and rates. Some of the ratios will have no units. It may be helpful for you to review Chapters 9 and 14 before proceeding further with this chapter.

Statistics and Averages

Approximately two thousand years ago, an aggressive emperor named Caesar Augustus insisted that all of the (his) world should be taxed. Therefore, all (male) persons were ordered to register in the town of their family's origin. History has noted that the best-remembered registrations were in Bethlehem, of a man named Joseph of Nazareth and his wife Mary. While they were in Bethlehem, Mary gave birth to a son.

Caesar was no fool. He knew that the taxes to be collected later did not always reach his Roman treasury. He, therefore, ordered an analysis of the registration data. (He wanted an estimate of how much tax should be collected and how much of that tax should be credited to Rome.) The person who performed this analysis became known as a **statistician.** This branch of mathematics became known as **statistics.**

The early statisticians developed measures known as **averages** to more simply describe a collection of related numbers. (Averages are only one of several measures used to describe large amounts of data.)

There are three types of averages:

- The **mean average** is often referred to as simply the **mean** or the **average.** You compute it by adding all of the related numbers together and then dividing that result by the **quantity of numbers** that you added together.
- The **median average** is often referred to as the **median.** It is the midpoint of a collection of related numbers arranged in **numerical order,** from the lowest number to the highest number.
- The **mode average** is often referred to as the **mode.** It is the one number that appears the most frequently in a collection of related numbers.

If the quantity of related numbers is small, then only the (mean) average or median is used to describe these numbers.

Additional measures are needed to more fully describe a collection of related numbers. A discussion of such measures can be found in specialized books on statistics and probability.

Study the following example. Observe the ways this collection of related numbers is described.

Example   Sixteen children attend a birthday party. The age of each child is recorded as the children arrive:

7, 5, 9, 3, 2, 4, 5, 5, 3, 7, 8, 5, 2, 5, 3, 7

Compute (a) the mean, (b) the median, and (c) the mode average ages.

Answer

a. The mean average age is

$$\frac{7 + 5 + 9 + 3 + 2 + 4 + 5 + 5 + 3 + 7 + 8 + 5 + 2 + 5 + 3 + 7}{16}$$

$$= \frac{80}{16}$$

$$= 5 \quad \text{the (mean) } \textbf{average } \text{age}$$

b. To compute the median average age, arrange the children's ages in numerical order:

2, 2, 3, 3, 3, 4, 5, 5, 5, 5, 5, 7, 7, 7, 8, 9

↑

The **Median** Age

We determined the median average age by counting from left to right until the space between the eighth and ninth numbers was reached. This location is the midpoint for this collection of related numbers. Because the midpoint is between two **fives,** the median age is **five.**

c. Another way of arranging these ages is to place identical ages in a vertical column. Then arrange these vertical columns in numerical sequence:

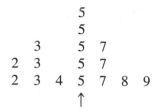

The **Mode** Age

The mode is easily determined; it is the age whose column is the highest **(5)**.

This type of arrangement can be boxed in as shown in the following figure:

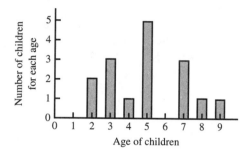

This type of display, known as a **histogram** or **bar chart,** will be discussed in more detail later in this chapter.

We may also display this information using a **pie chart.** A pie chart is a circle that has been divided into equal slices. For the preceding example, the circle could be divided into sixteen equal slices. Then the slices are grouped to represent the number of children for each age:

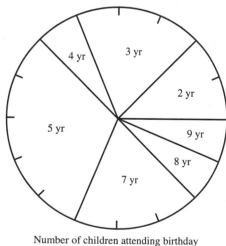

Number of children attending birthday
party grouped by age

Exercise Set 18.1

The answers to the exercises in this chapter are given in Appendix A.

Work the following exercises. Reduce all answers to either integers or decimal numbers.

1. A street contains nine houses. The prices of the nine houses are:

   $40 000; $60 000; $215 000, $40 000; $60 000;
   $40 000; $107 000; $40 000; $55 000

   a. Determine the mean, median, and mode.

   b. Using the following figure, prepare a histogram of these house prices:

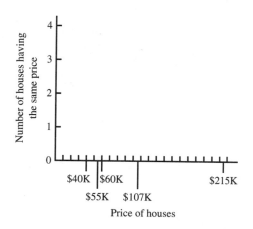

2. Total rainfall per week is measured for thirteen weeks (one quarter of a year). The results, to the nearest centimeter, are

   1, 0, 3, 2, 0, 2, 5, 4, 2, 1, 3, 2, 4

   a. Using the following figure, display this information as a bar chart:

AVERAGES AND PERCENT

b. List the thirteen values of rainfall in numerical order; then determine the three averages.

c. Using the following figure, prepare a histogram (bar chart) of weekly rainfall:

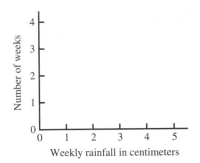

Weekly rainfall in centimeters

3. A study is made of tree growth. Twenty trees are planted. At the end of two years of growth, seventeen trees have survived. Their heights, to the nearest centimeter, are

16, 21, 13, 15, 18, 21, 15, 14, 22,
23, 15, 20, 17, 15, 17, 21, 23

a. List these seventeen values of tree height in numerical order. Then determine the three averages (based upon the number of living trees).

b. Make a bar chart; label the vertical axis **"Number of Living Trees Having the Same Height"** and the horizontal axis **"Tree Height in Centimeters."**

4. A study is made of bus use. The number of people in each bus are counted and recorded. The results from the counting of sixteen buses carrying people are

4, 7, 18, 16, 14, 5, 17, 15, 16,
17, 14, 16, 12, 9, 5, 2

a. List these sixteen values of **passenger count** in numerical order. Then determine the three averages.

b. Make a bar chart; label the vertical axis **"Number of Buses"** and the horizontal axis **"Number of Passengers in Each Bus."**

Percent

Another means of quantity comparison was developed for larger quantities of things. For example, assume that we have a package that contains one hundred buttons. The buttons are grouped by color as follows:

> 24 red buttons
> 32 orange buttons
> 17 green buttons
> 12 blue buttons
> 15 brown buttons
> 100 total buttons of all five colors

Let's compare the number of buttons of each color to the total number of buttons (100). We may then state that the buttons of each color are distributed as a portion of the one hundred total buttons. This portion is known as **percent,** meaning **per hundred.** Therefore, there are

> 24% red buttons
> 32% orange buttons
> 17% green buttons
> 12% blue buttons
> 15% brown buttons

Note: The % symbol is spoken **"percent."**

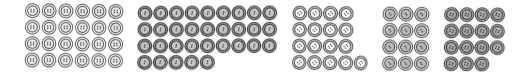

Let's examine another package of buttons. This next package contains

26 red buttons
43 orange buttons
29 green buttons
34 blue buttons
58 brown buttons
190 total buttons of all five colors

How can we more easily describe the distribution of these buttons? We can continue to use a **per hundred** or **percent** approach. The necessary formula is

$$\textbf{Percent} = \frac{\text{quantity of one type}}{\text{total quantity}} \times 100$$

For this package which contains a total of 190 buttons, the **percentage** of each color of button is as follows:

$$\frac{26}{190} \times 100 = 13.68\% \text{ red buttons}$$

$$\frac{43}{190} \times 100 = 22.63\% \text{ orange buttons}$$

$$\frac{29}{190} \times 100 = 15.26\% \text{ green buttons}$$

$$\frac{34}{190} \times 100 = 17.89\% \text{ blue buttons}$$

$$\frac{58}{190} \times 100 = 30.53\% \text{ brown buttons}$$

Note that the total of these five **percent** calculations is 99.99%, rather than 100%. Calculator approximations will often result in answers that are not quite exact, similar to a "definite maybe."

Exercise Set 18.2

Work the following exercises.

1. A package of buttons contains the following:

    39 red buttons
    72 orange buttons
    53 yellow buttons
    64 green buttons
    25 blue buttons
    46 brown buttons

    a. Compute the percentage of each color of button in the package.

    b. Add all six percentages together. Do they total 100%, or a number close to 100%?

AVERAGES AND PERCENT

2. A package of buttons contains the following:

27 red buttons
42 orange buttons
61 yellow buttons
14 green buttons
33 blue buttons
55 brown buttons

a. Compute the percentage of each color of button in the package.

b. Add all six percentages together. Do they total 100%, or a number close to 100%?

Histograms

Many histograms (or bar charts) are expressed using percentages instead of quantities along the vertical axis. Study the following example.

Example   Let's reconsider our earlier example involving sixteen children at a party, where the age of each child is as follows:

7,   5,   9,   3,   2,   4,   5,   5,   3,   7,   8,   5,   2,   5,   3,   7

a. Determine the percentage of children for each year of age.

b. Display these percentages as a histogram.

Answer

a. The percentages are given in the following table:

Years of Age	Number of Children	Percentage of Total Attendance
1	0	0
2	2	12.5
3	3	18.75
4	1	6.25
5	5	31.25
6	0	0
7	3	18.75
8	1	6.25
9	1	6.25

b. These percentages are displayed as a histogram in the following figure:

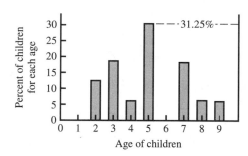

Compare this bar chart to the bar chart on page 18–5:

The bar chart vertical axis on page 18–5 is labeled

Number of Children for Each Age

The bar chart vertical axis in this example is labeled

Percent of Children for Each Age

The two bar charts have the same horizontal axis label.

If more than one number in a related group of numbers occurs the same (maximum) number of times, then there is more than one mode average. This group of numbers is said to have a **multimodal average.** Such a distribution is said to be **multimodal.**

Exercise Set 18.3

Work the following exercises.

1. Return to Exercise 1 of Exercise Set 18.1, which discussed the street containing nine houses priced as follows:

    $40 000, $60 000, $215 000, $40 000, $60 000,
        $40 000, $107 000, $40 000, $55 000

    a. Compute the percent of each price of house available.

    b. Construct a bar chart whose vertical axis is labeled **"Percent of Houses Having the Same Price"** and whose horizontal axis is labeled **"Price of Houses."**

2. Return to the tree growth study in Exercise 3 of Exercise Set 18.1, where the tree heights were as follows:

    16,   21,   13,   15,   18,   21,   15,   14,   22,   23,
        15,   20,   17,   15,   17,   21,   23

    a. Compute the percent of living trees that have the same height.

    b. What percentage of trees lived? _____

    c. What percentage of trees died? _____

3. Construct a bar chart whose vertical axis label is **"Percent of Living Trees Having the Same Height"** and whose horizontal axis label is **"Tree Height in Centimeters."**

4. Return to the button-counting example on page 18–9, where the buttons are grouped by color as follows:

$$\begin{array}{rl}
24 & \text{red buttons} \\
32 & \text{orange buttons} \\
17 & \text{green buttons} \\
12 & \text{blue buttons} \\
\underline{15} & \text{brown buttons} \\
100 & \text{total buttons of all five colors}
\end{array}$$

Display this information on a bar chart whose vertical axis label is **"Percent of Buttons Having the Same Color"** and whose horizontal axis label is **"Red   Orange   Yellow   Green   Blue   Brown."**

Work the following exercise involving a multimodal distribution.

5. On a street in a suburban neighborhood, the number of children per one-family house is

2, 5, 3, 2, 4, 6, 8, 7, 3, 5, 4, 3, 1, 2, 3, 1, 2

a. Determine the mean, median, and (two) modes for the number of children per family on this street.

b. Using the following figure, prepare a bar chart showing this distribution:

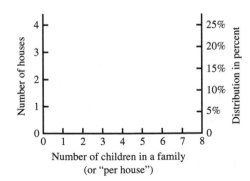

Displaying Mean Averages and Using Sigma (Σ) Notation

Mean averages can be determined from, and displayed on, another type of graph. The procedure for determining the mean average from this type of graph is to

- split the graph into rectangular pieces,
- compute the area of each rectangular piece,
- determine the sum of all areas (total area), and
- divide the total area by the distance along the **base.**

The result is the mean average of the height of the curve (graph) distribution.

Study the following example.

Example   A boy decides to mow lawns. He opens a bank account with $20 on week **0.** Then, on a weekly basis, he deposits his earnings (after expenses have been paid) into this bank account. At the end of his twenty-week season, he constructs the following graph of **Dollars Deposited per Week** (vertical axis) versus **Week of Deposit** (horizontal axis):

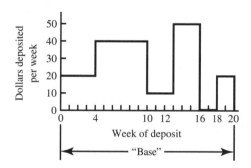

a. Compute the total amount deposited, which is the area between the curve (graph) and the horizontal axis.

b. Compute the average weekly deposit.

Answer

a. The formula for the area for each of the rectangles in this graph is

$$\text{Area} = \text{base} \times \text{height}$$

The **total area under the curve** (weeks × dollars/week) is computed as follows, where the Greek letter Σ (**sigma**) means "sum of":

$$\Sigma \text{ Area} = 4 \times 20 + 6 \times 40 + 2 \times 10 + 4 \times 50 + 2 \times 0 + 2 \times 20$$
$$= 80 + 240 + 20 + 200 + 0 + 40$$
$$= 580$$

Thus, the total of the deposits is $580.

b. The average weekly deposit is

$$\text{Average weekly deposit} = \frac{\Sigma \text{ area}}{\Sigma \text{ base}}$$

$$= \frac{\$580}{20 \text{ weeks}}$$

$$= \$29/\text{week}$$

This method of area computation was first suggested by Archimedes in 250 B.C. It was used by Sir Isaac Newton and Gottfried Wilhelm Leibniz in approximately A.D. 1670 when they separately developed the **integral calculus.**

Averages may be displayed as a straight, horizontal line through the original curve. The mean average value of the curve in the following figure (from the preceding example) is a horizontal line at **$29:**

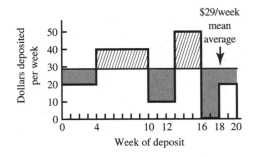

The sum of the areas above this line equals the sum of the areas below the line. As shown in Exercise 1 of page 18–19, withdrawals are indicated as lines below the horizontal axis. The areas between these lines and the horizontal axis are negative areas. Therefore, the areas below the horizontal axis are subtracted from the areas above the horizontal axis. Using math terminology, the **total area under the curve** is the difference between the absolute values of these two areas.

Exercise Set 18.4

Work the following exercise.

1. As in the example in the text, a boy mows lawns for twenty weeks. He starts a new bank account with a first deposit of $20. Twice during the season, he has to withdraw money to meet unexpected expenses. Examine the following graph:

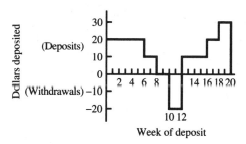

   a. Compute the total amount deposited.

   b. Compute the average weekly deposit.

   c. Display this average weekly deposit as a horizontal line on the given graph.

AVERAGES AND PERCENT

Summary

Averages permit us to look at fewer numbers and still obtain needed information. The properly chosen average provides us with a **feeling** for the total data. With averages, we do not have to examine each piece of that data. Consider the following example.

Example   Thirteen persons were asked to count the number of coins in their pockets. The results were as follows:

```
                6
  2   3     5  6  7
  2   3  4  5  6  7      9
```

The distribution of these coins may be displayed as a **histogram** or **bar chart.**

We determine the **(mean) average** number of coins per person by adding the total number of coins (**65**) divided by the number of persons (**13**); the result is **5.**

The **median (average)** is the midpoint of the number of coins, when they are numerically ordered from the smallest number of coins (**2**) to the largest number of coins (**9**). The midpoint, or median, is **5.**

The **mode (average)** is the one number of coins that occurs most frequently: **6.** Sometimes there is more than one mode. When there is more than one mode, then the distribution is said to be a **multimodal** distribution.

When there are a large number of items to be compared, then the **percent (%),** or per hundred, of each type of item is often computed, where

$$\text{Percent of one type} = \frac{\text{number of that type}}{\text{total number of items}} \times 100$$

For example, if there are **74** pennies with the date **1980** in a bag of **362** pennies, then the percent of 1980 pennies is

$$\text{Percent of 1980 pennies} = \frac{74}{362} \times 100$$
$$= 20.44\%$$

The study of distributions and their comparisons is known as **statistics.**

AVERAGES AND PERCENT

There are times when it is useful to determine and display the average of collected data. For example, study the following figure:

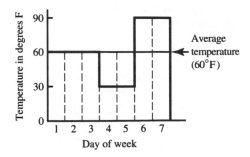

The (mean) average temperature for this week is

$$\text{Average °F} = \frac{\Sigma \text{ area}}{\Sigma \text{ base}}$$

$$= \frac{60 + 60 + 60 + 30 + 30 + 90 + 90}{7}\text{°F}$$

$$= \frac{420}{7}\text{°F}$$

$$= 60\text{°F}$$

Glossary of Math Words

Average (av′-er-ij) A nontechnical term that refers to the **mean average.**

Bar chart (bar chart′) A display consisting of separated (vertical) lines, or **bars,** where the height of each bar indicates the quantity or percent of the items being displayed; it is a simplified form of a **histogram.**

Histogram (his′-tuh-gram) A display consisting of a set of adjacent rectangles; the height of each rectangle indicates the actual quantity (or percent) of the items being displayed.

Mean (meen′) **average** The average that is computed by adding all of the separate related numbers together and then dividing by the total "number of numbers" that were added together. It is also the area **under** a curve divided by the base of that curve. The **mean average** is also known as simply the **average.**

Median (mee′-dee-an) **average** The midpoint of a collection of numbers (or objects) that have been arranged in numerical order (or worth) from the lowest to the highest.

Mode (mowed′) **average** The number, or type, of objects that occurs most frequently in a collection of numbers or objects. If more than one number or object occurs the same (maximum) number of times, then that distribution is **multimodal.**

Multimodal (mul′ti-mow′-dal) **average** *See* Mode average.

Percent (per-sent′)   The distribution of a number of units as a portion of one hundred.

Pie (pye′) **chart**   A circle that has been split into pieces; the size of each **piece of pie** is chosen to represent the desired portion of the **entire pie.**

Sigma (sig′-ma)   The Greek letter Σ which indicates that a sum is to be determined.

Statistics (stuh-tiss′-tiks)   The branch of mathematics where collections of numbers are classified and analyzed; the study of distributions and their comparisons.

Chapter 18 Test

Follow instructions carefully:

> *On a separate piece of paper, write the answers to the following questions. Do* not *write on these pages.*
>
> *When you are finished, compare your answers with those given in Appendix B.*
>
> *Record the date, your test time, and your score on the chart at the end of this test.*

Seven families consist of the following number of children:

1,   2,   5,   2,   9,   6,   3

1.   The mean average number of children per family is _____.

2.   The median average is _____.

3.   The mode average is _____.

A girl mows lawns during her ten-week vacation. She deposits the following amounts each week in her savings account:

$10	$40
$30	$20 withdrawal
$20	$10
$30	$10
$10	$20

4.   Construct a graph of **"Dollars Deposited"** (vertical axis) versus **"Week of Deposit"** (horizontal axis).

5.   The girl's average weekly deposit is _____.

A box of buttons is emptied and counted. The results are

> 27 green
> 36 red
> 49 yellow
> 18 blue

6. The percent of red buttons is _____.

7. The percent of blue buttons is _____.

8. The percent of yellow buttons is _____.

9. The percent of green buttons is _____.

10. Construct a bar chart that indicates the distribution of these buttons. Label the vertical axis **"Percent of Total Buttons."** Label the horizontal axis **"Color of Button."**

11. An average of $7/week is deposited in a savings account for twelve weeks. The total of the deposits is _____.

The lowest daily temperatures in Methuen, Massachusetts, recorded over a period of two consecutive weeks, were as follows:

> 23°F, 17°F, 38°F, 28°F, 15°F, 19°F, 28°F,
> 35°F, 42°F, 21°F, 23°F, 28°F, 40°F, 35°F

12. Determine the mean daily temperature in Methuen for these two weeks.

13. Determine the median of the daily temperatures in Methuen for these two weeks.

14. Determine the mode of the daily temperatures in Methuen for these two weeks.

15. Sketch a bar chart applying the preceding data.

16. What is the percent of days whose lowest temperature is 23°F?

17. What is the percent of days whose lowest temperature is between 20°F and 30°F?

A girl deposits the following amount in her bank account each day:

$$\$10, \quad \$30, \quad -\$10 \text{ (a withdrawal)}, \quad \$20, \quad \$20, \quad \$10$$

18. Construct a graph of **"Dollars Deposited"** (vertical axis) versus **"Day of Deposit"** (horizontal axis).

19. The girl's average daily deposit is _____.

20. Three people each earn $225 per week, five people each earn $325 per week, and four persons each earn $475 per week. Determine the average salary earned per week.

Chapter 18 Test Record

DATE	TIME	SCORE

Refer to

- Appendix B for the correct answers to this test.
- Appendix C if taking this test required too much effort.
- Appendix D for **Memory Methods** assistance.
- Appendix E if your test scores are decreasing.

When a problem seems difficult, find one like it in this chapter. Then study that (and the related material) again.

Develop additional 3 × 5 cards for those ideas, problems, and procedures that caused you difficulty.

Slopes and Rates

Slope and Change

Comparisons are often made between two numbers. Two types of comparisons that we have explored are the difference between two numbers and the ratio of two numbers. In this chapter, we shall explore **slopes** and **rates;** each is the ratio of two numbers. These ratios are used to describe **change.**

These ratios are often used in music, science, engineering, and finance; they are also used to describe the motion of objects.

Comparisons that involve ratios often include units of measure. If the same unit of measure occurs in both the numerator and the denominator of a ratio, then these units will cancel. (Refer to Chapter 14 of Book 1, where we discussed units of measure.)

Study the following example.

Example

$$\frac{3 \text{ meters}}{4 \text{ meters}} = \frac{3}{4} = 0.75$$

The resulting value of this ratio, **0.75,** is referred to as a **unitless ratio** because no units remain; they have been canceled.

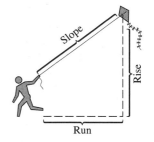

A change in vertical distance is often referred to as a **rise.** A change in horizontal distance is often referred to as a **run.**

Slope is defined as the ratio of these two types of change:

$$\text{Slope} = \frac{\text{change in vertical distance}}{\text{change in horizontal distance}} = \frac{\text{rise}}{\text{run}}$$

This ratio is often a unitless ratio.

Study the following example.

Example   What is the slope (change) of a line whose vertical distance is **3** and whose corresponding horizontal distance is **5?**

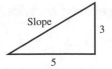

Answer

$$\text{Slope} = \frac{3}{5} = 0.6$$

Many roads are not horizontal; they **slope.** However, the slope is so gradual that the slope is usually given in **percent.** When the slope of a road is given in percent, then the slope is known as the **grade** of that road.

Study the following example.

Example   A road rises 2 feet for every 50 feet of horizontal distance. What is the grade of this road?

Answer

$$\text{Grade} = (\text{slope}) \times (100)$$
$$= \frac{2}{50} \times 100$$
$$= 4\%$$

We read from left to right. Therefore, the sign of slope is defined from left to right. For example:

A road that rises from left to right is said to have a **positive slope.**

A road that falls from left to right is said to have a **negative slope.**

Exercise Set 19.1

The answers to the exercises in this chapter are given in Appendix A.

Work the following exercises.

1. Determine the slope of a line whose rise is 4 feet and whose run is 8 feet.

2. What is the slope of a line whose rise is 352 centimeters and whose run is 5.6 meters?

3. Determine the grade of a road whose rise is 75 centimeters and whose run is 40 meters.

4. Determine the grade of a road whose drop is 2 feet and whose run is 80 feet.

Review of the Distance-Subtraction Rule

In Chapters 9 and 10 of Book 1, we explored horizontal and vertical real number lines. We also examined the length of bars (scalars) along these lines. We determined the distance along the horizontal real number line by subtracting the left distance value from the right distance value.

For example, on the following real number line, the **horizontal distance** between **30** and **−20** is

$$30 - (-20)$$
$$= 30 + 20$$
$$= 50$$

We also determined distance along the vertical real number line by subtracting the bottom distance value from the top distance value.

For example, in the following figure, the **vertical distance** between **40** and **−30** is

$$40 - (-30)$$
$$= 40 + 30$$
$$= 70$$

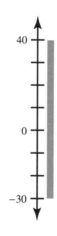

Exercise Set 19.2

Work the following exercises.

1. Determine the distance along the horizontal of the following real number line:

2. Determine the distance along the vertical of the following real number line:

3. Determine the distance along the horizontal of the following real number line:

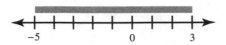

4. Determine the distance along the vertical of the following real number line:

The Cartesian Coordinate System

In approximately A.D. 1620, the pilgrims were arriving on land now known as Massachusetts. In France, René Descartes was devising a system for constructing graphs. His **Cartesian coordinate system** simplified computations involving measure in two and three dimensions. This system has been of great value to many people, including navigators and mathematicians.

René Descartes combined the horizontal and vertical real number lines in his new coordinate system. He named the horizontal (east, west) axis *x* and the vertical (north, south) axis *y*.

He recommended that a **point location** in this system would always be written in the following sequence:

- an initial parenthesis,
- the horizontal distance from the origin, followed by a comma,
- the vertical distance from the origin, and
- an ending parenthesis.

Study the following examples.

Example Locate point *A* at (2, 3) and point *B* at (−4, −5).

Answer Note in the following figure that the *x*- and *y*-axes share the point (0, 0), the **origin.** Observe points *A* and *B* shown on the *x*- and *y*-axes.

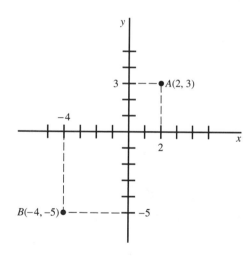

We determine the horizontal (x) distance between A and B as follows:

$$2 - (-4)$$
$$= 2 + 4$$
$$= 6$$

We determine the vertical (y) distance between A and B as follows:

$$3 - (-5)$$
$$= 3 + 5$$
$$= 8$$

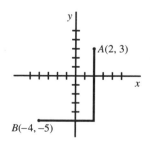

Example Determine the slope of the line from point *A* to point *B* in the following figure. This is the **slant line.**

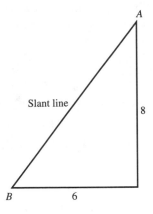

Answer

$$\text{Slope} = \frac{\text{rise}}{\text{run}}$$

$$= \frac{8}{6}$$

$$= 1.\overline{3}$$

A line in the Cartesian coordinate system may have one of four types of slope:

Four Types of Slope

1. A line that starts in the lower left of a coordinate system and ends in the upper right of the same system has a positive (+) slope.

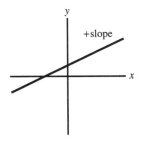

2. A line that starts in the upper left of a coordinate system and ends in the lower right of that same system has a negative (−) slope.

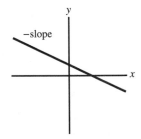

3. A line that has no (vertical) rise has a zero (0) slope.

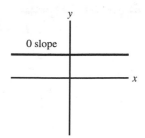

4. A line that has no (horizontal) run has an infinite (∞), or undefined, slope.

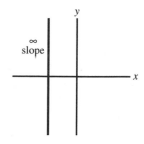

Recall that division by zero has an **undefined** or infinite (∞) result.

Study the following examples.

Example   Compute the slope of a line between points $A(8, 2)$ and $B(3, -1)$, as shown in the following figure:

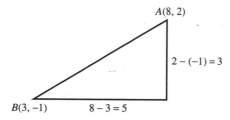

Answer

$$\text{Slope} = \frac{3}{5}$$
$$= 0.6 \quad \text{or} \quad +0.6$$

Example Compute the slope of a line between points $A(-2, 5)$ and $B(2, 3)$, as shown in the following figure:

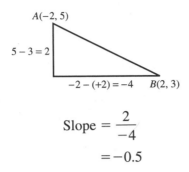

$A(-2, 5)$

$5 - 3 = 2$

$-2 - (+2) = -4$ $B(2, 3)$

$$\text{Slope} = \frac{2}{-4}$$

$$= -0.5$$

Point	x-Value	y-Value
A	-2	5
B	2	3
$A - B$	-4	2

Note that you may also subtract the x- and y-coordinate value of point B from point A, as shown in the following example.

Example We may compute the value of the slope in our first example as follows:

$$\text{Slope} = \frac{3 - 5}{2 - (-2)}$$

$$= \frac{-2}{4}$$

$$= -0.5$$

The slope value remains the same.

However, note from the previous example that you will compute an incorrect slope value if you subtract B from A for the numerator and A from B for the denominator. Similarly, you will have an incorrect slope value if you subtract A from B for the numerator and B from A for the denominator.

You may verify that you have the correct sign of the slope value for a line in the Cartesian coordinate system. Compare the computed slope values of the previous examples with the four types of slope described on pages 19–10 and 19–11. Then study the following examples.

Example   Compute the slope of a line between points $A(-3, 2)$ and $B(4, 2)$, as shown in the following figure:

$A(-3, 2)$ $B(4, 2)$

$4 - (-3) = 7$

Answer   Note that there is no vertical change $(0 - 0 = 0)$. Thus,

$$\text{Slope} = \frac{0}{7}$$
$$= 0$$

Example   Compute the slope of a line between points $A(3, 2)$ and $B(3, -4)$.

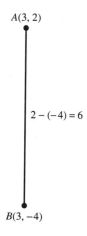

$A(3, 2)$

$2 - (-4) = 6$

$B(3, -4)$

Answer   Note that there is no horizontal change $(0 - 0 = 0)$. Thus,

$$\text{Slope} = \frac{6}{0}$$
$$= \infty \quad \text{or} \quad \text{undefined}$$

Exercise Set 19.3

Work the following exercises.

1. Construct an *x*-axis and a *y*-axis through the common origin.

    a. Plot points $A(5, 8)$ and $B(-7, 3)$.

    b. The horizontal distance from *A* to *B* is _____.

    c. The vertical distance from *A* to *B* is _____.

    d. The slope of the slant line from *A* to *B* is _____.

2. Construct an *x*-axis and a *y*-axis through the common origin.

    a. Plot points $A(-3, 2)$ and $B(1, -4)$.

    b. The horizontal distance between *A* and *B* is _____.

    c. The vertical distance between *A* and *B* is _____.

    d. The slope of the line from *A* to *B* is _____.

3. Construct an *x*-axis and a *y*-axis through the common origin.

    a. Plot points $A(-3, 5)$ and $B(-3, -7)$.

    b. The horizontal distance between *A* and *B* is _____.

    c. The vertical distance between *A* and *B* is _____.

    d. The slope of the line from *A* to *B* is _____.

4. Construct an *x*-axis and a *y*-axis through the common origin.

    a. Plot points $A(-7, 6)$ and $B(3, 6)$.

    b. The horizontal distance between *A* and *B* is _____.

    c. The vertical distance between *A* and *B* is _____.

    d. The slope of the line from *A* to *B* is _____.

5. Construct an *x*-axis and a *y*-axis through the common origin.

    a. Plot points $A(-2, 7)$ and $B(3, -2)$.

    b. The horizontal distance between *A* and *B* is _____.

    c. The vertical distance between *A* and *B* is _____.

    d. The slope of the line from *A* to *B* is _____.

6. Construct an *x*-axis and a *y*-axis through the common origin.

    a. Plot points $A(-5, 4)$ and $B(-1, -4)$.

    b. The horizontal distance between *A* and *B* is _____.

    c. The vertical distance between *A* and *B* is _____.

    d. The slope of the line from *A* to *B* is _____.

A special math symbol is used to indicate a difference along a line. The Greek letter **delta** (Δ) means a "change" or "difference":

When a distance along the x-axis is being computed, then the distance may be indicated as Δx (spoken "**delta x**"). For example:

$$\begin{aligned} \Delta x &= 30 - (-20) \\ &= 30 + 20 \\ &= 50 \end{aligned}$$

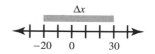

When a distance along the y-axis is being computed, then the distance may be indicated as Δy (spoken "**delta y**"). For example:

$$\begin{aligned} \Delta y &= 20 - (-30) \\ &= 20 + 30 \\ &= 50 \end{aligned}$$

Now we may examine a more mathematical definition of **slope:**

$$\text{Slope} = \frac{\Delta y}{\Delta x}$$

The slope of a slant line, with these values of Δx and Δy, is

$$\text{Slope} = \frac{\Delta y}{\Delta x} = \frac{50}{50} = 1$$

Remember:

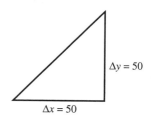

Δx means **a change along the x-axis.**
Δy means **a change along the y-axis.**

Slant Line Distance: The Pythagorean Theorem

In a large number of situations, it is necessary to compute the slant line distance between two points. For many centuries, the formula eluded **geometers** and **mathematicians.** In approximately 530 B.C., Pythagoras the Phoenician solved this math problem. (As a person with math anxiety, he wasn't sure that he had solved it for all possibilities; he had!) He studied **triangles** that have one horizontal side (Δx) and one vertical side (Δy). (Recall from Chapter 17 that these triangles are also known as **right triangles.**) He proved that the slant side (Δs), or the **hypotenuse,** has a length that is always

$$\Delta s = \sqrt{(\Delta x)^2 + (\Delta y)^2}$$

The Pythagorean theorem applies only to right triangles.

This formula is known as the **Pythagorean theorem.** The only restriction is that the units of measure of **all three sides** must be the same. The following figure illustrates the sides of a right triangle:

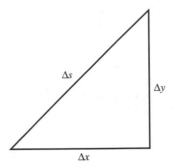

In 530 B.C., mathematicians could perform only those computations that involved integers. A **square (x^2) key** and a **square root ($\sqrt{\ }$) key** are provided on modern hand-held calculators. We may, therefore, determine the slant side distance (Δs) for any numbers that might be of interest to us.

Study the following examples.

Example   A right triangle has a 30-cm horizontal side and a 40-cm vertical side as shown in the following figure:

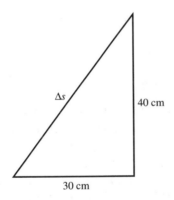

Determine the length of the hypotenuse (Δs).

Answer

$$\Delta s = \sqrt{(30)^2 + (40)^2} \text{ cm}$$
$$= \sqrt{900 + 1600} \text{ cm}$$
$$= \sqrt{2500} \text{ cm}$$
$$= 50 \text{ cm}$$

Example   A right triangle has a 25-in. horizontal side and a 36-in. vertical
side as shown in the following figure:

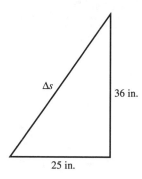

Determine the length of the hypotenuse (Δs).

Answer

$$\Delta s = \sqrt{(25)^2 + (36)^2} \text{ in.}$$
$$= \sqrt{625 + 1296} \text{ in.}$$
$$= \sqrt{1921} \text{ in.}$$
$$= 43.829 \text{ in.}$$

Exercise Set 19.4

Work the following exercises.

1. A right triangle has a 6-in. horizontal side and an 8-in. vertical side.

   a. Determine the length of the hypotenuse (Δs).

   b. Construct the right triangle and indicate the three dimensions on that triangle.

2. A right triangle has a 13.6-cm horizontal side and a 9.4-cm vertical side.

   a. Determine the length of the hypotenuse (Δs).

   b. Construct the right triangle and indicate the three dimensions on that triangle.

3. A right triangle has a 5-cm horizontal side and a 12-cm vertical side.

   a. Determine the length of the hypotenuse (Δs).

   b. Construct the right triangle and indicate the three dimensions on that triangle.

   c. Determine the slope of the hypotenuse.

4. A right triangle has a 32.4-in. horizontal side and a 21.8-in. vertical side.

    a. Determine the length of the hypotenuse.

    b. Construct the right triangle and indicate the three dimensions on that triangle.

    c. Determine the slope of the hypotenuse.

5. Construct an x-axis and a y-axis through the common origin.

    a. Plot points $A(-4, 5)$ and $B(3, -6)$.

    b. The horizontal distance between A and B is _____.

    c. The vertical distance between A and B is _____.

    d. The slope of the line from A to B is _____.

    e. The slant line distance from A to B is _____.

6. Construct an x-axis and a y-axis through the common origin.

    a. Plot points $A(3, 7)$ and $B(-6, 2)$.

    b. The horizontal distance between A and B is _____.

    c. The vertical distance between A and B is _____.

    d. The slope of the line from A to B is _____.

    e. The slant line distance from A to B is _____.

René Descartes developed the Cartesian coordinate system in approximately 1640; he used the Pythagorean theorem to compute slant line distances. The next extension of his coordinate system was to allow for computations involving the third dimension: **depth,** noted as Δz. The formula for computing a three-dimensional slant line distance (Δs) is a simple extension of the formula of Pythagoras:

$$\Delta s = \sqrt{(\Delta x)^2 + (\Delta y)^2 + (\Delta z)^2}$$

Study the following example.

Example   Compute the three-dimensional, slant line distance (Δs) in the following figure, where

$$\Delta x = 7 \qquad \Delta y = 5 \quad \text{and} \quad \Delta z = 4$$

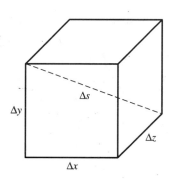

Answer

$$\Delta s = \sqrt{(7)^2 + (5)^2 + (-4)^2}$$
$$= \sqrt{+49 + 25 + 16}$$
$$= \sqrt{+90}$$
$$= 9.487$$

Exercise Set 19.5

Work the following exercises.

1. Compute the three-dimensional, slant line distance, where $\Delta x = 5$, $\Delta y = 2$, and $\Delta z = 4$.

2. Compute the three-dimensional, slant line distance, where $\Delta x = 8.2$ cm, $\Delta y = -5.6$ cm, and $\Delta z = 9.3$ cm.

3. Compute the three-dimensional, slant line distance, where $\Delta x = -7$, $\Delta y = 5$, and $\Delta z = -4$.

4. Compute the three-dimensional, slant line distance, where $\Delta x = -9.6$ cm, $\Delta y = -7.8$ cm, and $\Delta z = 6.7$ cm.

Measuring Motion

As we walk, cycle, drive, or fly from house to house, city to city, nation to nation, Earth to moon, or throughout the universe, we want answers to such questions as:

> Where have we been, and how far have we traveled? Where will we be a minute, hour, or day from now?
>
> How fast are we traveling? How fast could we travel?
>
> How quickly could we move from **rest** to **full speed?** How quickly could we stop in an emergency?
>
> How long has it taken us? How long will it take us?

These four sets of questions either directly involve the motion of an object or are the result of that object's motion.

Motion involves time. We measure time in fractions of a second, seconds, minutes, hours, days, and years. Scientists consider **time** to be the **fourth dimension** of measure. Time is used to explain how we, or our vehicles, are first in one place and then in another. Time is used to explain **motion.**

Sit in a chair. Tell another person to move to a different location while you close your eyes. Because that person moved while your eyes were closed, you did not see the motion. You have to assume that while time elapsed (changed), the person was moving.

Assume that you closed your eyes for ten seconds (Δt) and the person moved a distance of 20 feet (Δs). The **average speed** of that person's motion is defined as the change in distance (Δs) divided by the change in time (Δt):

$$\text{Average speed} = \frac{\Delta s}{\Delta t} = \frac{20 \text{ ft}}{10 \text{ sec}} = 2 \text{ ft/sec}$$

Notice that if the person moved from west to east 20 feet, then the **average velocity** of that person's motion was

$$\text{Average velocity} = \frac{\Delta s}{\Delta t} \quad \text{to the east}$$

$$= 2 \text{ ft/sec} \quad \text{to the east}$$

The person traveled at a **rate** of 2 ft/sec to the east.

Note that the word **rate** was used in the preceding sentence. The general term for a fraction is a **ratio** or **rate.** When the denominator of a fraction involves time, then this special ratio is known as either a **rate** or a **time-related rate.**

Note that rates do not always involve distance. The rate of interest we pay to borrow money might be expressed as

12% per year (annum)

Recall that time-related rates involve **time** in their denominators.

Velocity includes a value and a direction; velocity is a **vector** quantity. Speed includes only a value; speed is a **scalar** quantity. **Speed** and **velocity** describe a change in distance when compared to a change in time. Note that the word **speed** is sometimes used when **velocity** is the correct choice. The opposite is also true.

Example

I was caught *speeding* 72 miles/hour east. (*wrong*)
The *velocity* of my car was 30 km/h. (*also wrong*)

Recall from Chapter 11 that distance is used to describe both vector and scalar quantities:

When we compute slope, we use distance as a vector quantity.

When we compute a total distance traveled, we use distance as a scalar quantity. Distance traveled may involve one, two, or three dimensions.

Rate computations involving distance, time, and velocity are often presented with graphs. These graphs provide a visual presentation of the distance traveled and its companion, velocity.

Graphing Motion

Study the following examples.

Example   A car travels at a constant velocity from its starting (reference) point, or origin, to a distance 120 miles north in 3 hours. Determine the velocity of the car. Display graphically both the distance traveled and the velocity.

Answer   The velocity (v) of the car is

$$v = \frac{\Delta s}{\Delta t}$$

$$v = \frac{120 - 0}{3 - 0} \text{ mi/hr north}$$

$$= \frac{120}{3} \text{ mi/hr north}$$

$$= 40 \text{ mi/hr north}$$

The graphs of distance (*s*) versus time (*t*), and of velocity (*v*) versus time (*t*), are shown in the following figures for the preceding example:

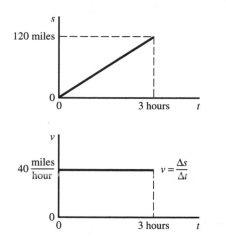

The graph that displays distance (Δs) on the vertical (*y*) axis and time (Δt) on the horizontal (*x*) axis is constructed first. This graph is known as a distance-versus-time graph.

Directly beneath this graph, the graph that displays velocity (*v*) on the vertical (*y*) axis and time (Δt) on the horizontal (*x*) axis is constructed. This graph is known as a velocity-versus-time graph.

Note that the velocity curve is a horizontal straight line. This line is horizontal because velocity is a constant. Its value is 40 miles per hour north for each moment of time from 0 to 3 hours.

This (horizontal) line is also the graph of the **slope** of the **distance** curve. In advanced math texts, it would be indicated as the **first derivative of the distance curve.** We will use the word **slope** because such expressions as **differential calculus** might be frightening to you!

Sometimes we want to compute the distance traveled. When the velocity (or speed) is constant, then the simplest formula for computing the distance traveled is

$$\Delta s = v \, \Delta t \quad \text{where the velocity } v \text{ is constant}$$

Study the example on the following page.

Example   A car leaves its starting point (origin) at 10:00 A.M. It travels at a constant velocity of 80 kilometers per hour east until 1:00 P.M. Construct the graphs of distance versus time and of velocity versus time.

Answer   Study the following detailed solution to this example.

The formula for distance is

$$\Delta s = v \, \Delta t$$
$$= (80 \text{ km/h east}) \times (\Delta t)$$

The total distance is traveled in three hours (1:00 P.M. minus 10:00 A.M.).

For $\Delta t = 0$ hours   (10:00 A.M. minus 10:00 A.M.),

$$\Delta s = (80 \text{ km/h east}) \times (0 \text{ h})$$
$$= 0 \text{ km}$$

For $\Delta t = 1$ hour   (11:00 A.M. minus 10:00 A.M.),

$$\Delta s = (80 \text{ km/h east}) \times (1 \text{ h})$$
$$= 80 \text{ km}$$

For $\Delta t = 2$ hours   (noon minus 10:00 A.M.),

$$\Delta s = (80 \text{ km/h east}) \times (2 \text{ h})$$
$$= 160 \text{ km}$$

For $\Delta t = 3$ hours   (1:00 P.M. minus 10:00 A.M.),

$$\Delta s = (80 \text{ km/h east}) \times (3 \text{ h})$$
$$= 240 \text{ km}$$

This information is used to determine the following table of values:

Time in Hours	Distance in Kilometers	Velocity in km/h
0	0	80 E
1	80	80 E
2	160	80 E
3	240	80 E

This table is also known as a **chart** or **matrix.**

These values are also used to construct the following graphs of distance versus time and of velocity versus time:

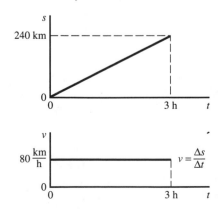

Exercise Set 19.6

Work the following exercises.

1. A bicyclist travels at a constant velocity from her starting point to a distance 40 kilometers east in four hours.

   a. Determine the velocity of the bicyclist.

   b. Construct the graphs of distance versus time and of velocity versus time.

2. A man leaves his starting point at 9:00 A.M. He walks at a constant velocity of 5 kilometers per hour to the north until 2:00 P.M.

    a. Using the proper formula for distance, determine the distance traveled.

    b. Construct the graphs of distance versus time and of velocity versus time.

There are situations where a moving object may start from a point that is not the origin. The distance formula must include another term. This term is the distance between the object's starting point and the origin.

This reference point at $t = 0$, known as the **initial value**, is usually labeled s_0. It is spoken "s sub-zero," where the zero is known as a **subscript.** A subscript is written lower than, and to the right of, a letter.

The modified formula for distance (s) is

$$s = vt + s_0$$

Note that we have written Δs as s because the distance s is measured from the origin. The total distance traveled, Δs, is equal to the value of s on the number line.

The s consists of the distance traveled, $v \, \Delta t$ or Δs, plus the distance from the origin to the starting point, s_0.

We have written Δt as t for the same reason that Δs can be written as s.

The t is the elapsed time starting at $t = 0$, as illustrated in the following figure:

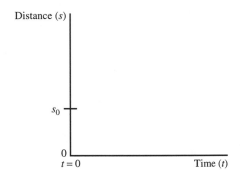

Also note that the value of s_0 is the distance measured from the origin to the object's location at $t = 0$.

Study the example on the following page.

Example   A runner starts from a point that is 50 meters east of the actual reference point. The runner moves at a constant velocity of 15 meters per second east for 30 seconds.

a. How far is the runner from the reference point after 30 seconds?

b. Construct the graphs of distance versus time and of velocity versus time.

Answer

a. $s = vt + s_0$
$= 15 \text{ m/s} \times 30 \text{ s} + 50 \text{ m}$
$= 450 \text{ m} + 50 \text{ m}$
$= 500 \text{ m}$

Therefore, the runner is 500 meters east of the reference point after 30 seconds.

b. The graphs are shown in the following figures:

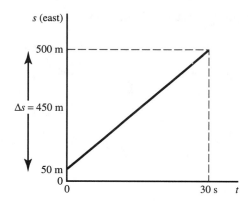

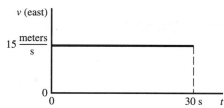

Exercise Set 19.7

Work the following exercises.

1. A runner starts from a point that is 40 meters north of the actual reference point. The runner moves at a constant velocity of 10 meters per second north for 20 seconds.

    a. How far is the runner from the reference point after 20 seconds?

    b. Construct the graphs of distance versus time and of velocity versus time.

2. A car starts driving at a constant velocity from a point 60 miles east of a reference point. Three hours later the car is 180 miles east of that same reference point.

    a. Construct the graph of distance versus time.

    b. Use the formula $v = \Delta s/\Delta t$ to determine the value of the constant velocity.

    c. Construct the graph of velocity versus time below the first graph.

Study the following example.

Example   A runner starts from a point that is 50 meters west of the actual reference point. The runner moves at a constant velocity of 15 meters per second east for 30 seconds.

a. How far is the runner from the reference point after 30 seconds?

b. Construct the graphs of distance versus time and of velocity versus time.

Answer

a. $s = vt + s_0$
$\quad = 15 \text{ m/s} \times 30 \text{ s} + (-50) \text{ m}$
$\quad = 450 \text{ m} - 50 \text{ m}$
$\quad = 400 \text{ m}$

Therefore, the runner is 400 meters east of the reference point after 30 seconds.

b. The graphs are shown in the following figures:

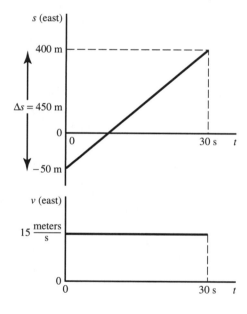

Exercise Set 19.8

Solve the following exercises.

1. A car starts driving at a constant velocity from a point 40 miles south of a reference point. Three hours later the car is 80 miles north of that same reference point.

   a. Construct the graph of distance versus time; north is shown as +, and south is shown as −.

   b. Use the formula $v = \Delta s/\Delta t$ to determine the value of the constant velocity.

   c. Construct the graph of velocity versus time below the first graph.

   d. Using the graph of distance versus time, estimate the time at which the car will pass the reference point.

SLOPES AND RATES

2. A car starts driving at a constant velocity from a point 50 miles south of a reference point. Four hours later the car is 250 miles north of that same reference point.

    a. Construct the graph of distance versus time; north is shown as +, and south is shown as −.

    b. Use the formula $v = \Delta s/\Delta t$ to determine the value of the constant velocity.

    c. Construct the graph of velocity versus time below the first graph.

    d. Using the graph of distance versus time, estimate the time at which the car will pass the reference point.

Acceleration

When we buy a car, we often want to know how fast the car can change its speed or velocity, that is, how fast it can **accelerate.**

Acceleration describes a change in velocity. If a car can change its speed (Δv) from 10 ft/sec to 40 ft/sec in 6 seconds (Δt), then the average acceleration of that car is

$$\text{Average acceleration} = \frac{\Delta v}{\Delta t}$$

$$= \frac{(40 - 10) \text{ ft/sec}}{6 \text{ sec}}$$

$$= \frac{30 \text{ ft/sec}}{6 \text{ sec}}$$

$$= (5 \text{ ft/sec})/\text{sec}$$

$$= 5 \text{ ft/sec}^2$$

The car can accelerate at the **rate** of 5 ft/sec². These units of acceleration are spoken "**feet per second per second**" or "**feet per second squared.**"

Velocity is a time-related rate. **Acceleration** is also a time-related rate; it is defined as a change in velocity (Δv), or speed, divided by a change in time (Δt).

Most cars are designed to develop a constant acceleration. The three simplified formulas that relate acceleration (a), changing velocity (v), and accumulating distance (s) are as follows:

$$s = \frac{1}{2}at^2$$

$$v = at$$
$$a = \text{constant}$$

Deceleration occurs whenever the velocity of a body is decreasing. It is the negative of acceleration.

Study the example on the following page.

Example A car accelerates from rest, at its reference point, with a constant acceleration of 2 m/s^2.

    a. Using the preceding formulas, compute the values of distance, velocity, and acceleration versus time from $t = 0$, every second, to $t = 5$ seconds.

    b. Construct the three graphs; the top graph should be distance, the middle graph should be velocity, and the bottom graph should be acceleration (versus time).

Answer The values and graphs are shown in the following figure:

t	s	v	a
0	0	0	2
1	1	2	2
2	4	4	2
3	9	6	2
4	16	8	2
5	25	10	2

This table is also known as a chart, or matrix. The values are used to construct the graphs.

    Note that the units of distance (s) are meters (m), the units of velocity (v) are meters per second (m/s), and the units of acceleration (a) are meters per second squared (m/s^2).

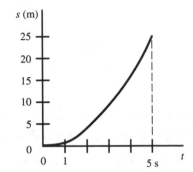

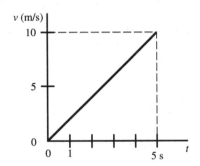

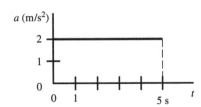

Exercise Set 19.9

Work the following exercises.

1. A car accelerates from rest, from its reference point, at a constant value of 4 ft/sec^2.

   a. Compute the values of distance, velocity, and acceleration versus time from $t = 0$ to $t = 5$ seconds.

   b. Record the results of these computations using a matrix of data.

   c. Construct the three graphs. The top graph should be distance versus time, the middle graph should be velocity versus time, and the bottom graph should be acceleration versus time. Label the numerical values and units on each of the axes.

2. A car accelerates from rest, from its reference point, at a constant value of 6 ft/sec^2.

   a. Compute the values of distance, velocity, and acceleration versus time from $t = 0$ to $t = 5$ seconds.

   b. Record the results of these computations using a matrix of data.

   c. Construct the three graphs. The top graph should be distance versus time, the middle graph should be velocity versus time, and the bottom graph should be acceleration versus time. Label the numerical values and units on each of the axes.

MATH AT WORK

Summary

The **slope** of a curve is a ratio. Slopes may be displayed on a graph of vertical (*y*) values versus horizontal (*x*) values, as shown in the following figure:

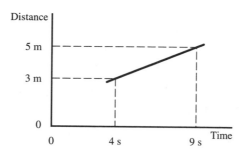

If the horizontal (*x*) axis is used to display time, then the ratio is referred to as either a **rate** or a **time-related rate.**

Slopes and rates are defined as the following ratio:

$$\text{Slope} = \frac{\text{change in vertical distance}}{\text{change in horizontal distance}}$$

$$= \frac{\text{rise}}{\text{run}} = \frac{2}{5} = 0.4$$

Road grades are also slopes. To determine the road grade, we multiply the slope of the road by 100:

$$\text{Grade} = (\text{slope}) \times (100)$$

The grade is expressed in percent:

$$\text{Grade} = 0.4 \times 100 = 4\%$$

Distances can be presented on a graph of vertical versus horizontal values using the **Cartesian coordinate system. Point locations** on a two-dimensional graph are written in a special way. The point coordinates are written within a pair of parentheses, with the *x*-value first, a comma, and the *y*-value last. A point whose coordinates are **5** units to the right of the origin, and **6** units down from the origin, is labeled as (5, −6), as shown in the following figure:

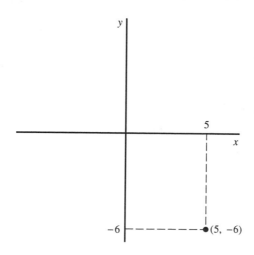

A line may have one of four types of slope:

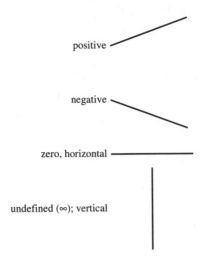

positive

negative

zero, horizontal

undefined (∞); vertical

The Greek letter **delta** (Δ) is used in math to indicate a difference. Therefore, the more common definition of slope uses the **delta** math symbol.

Study the following example.

Example Determine the slope of the line between the points (9, 5) and (4, 3).

Answer

$$\text{Slope} = \frac{\text{rise}}{\text{run}} = \frac{\Delta y}{\Delta x}$$

$$= \frac{5 - 3}{9 - 4}$$

$$= \frac{2}{5}$$

$$= 0.4$$

The slope of the slant line is 0.4, as shown in the following figure:

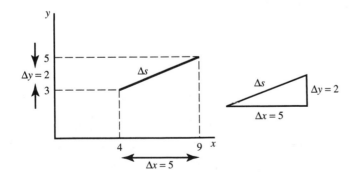

The distance along a slant line can be computed using the **Pythagorean theorem.** For a two-dimensional line, the distance along the slant line (Δs) is

$$\Delta s = \sqrt{(\Delta x)^2 + (\Delta y)^2} = \sqrt{(5)^2 + (2)^2}$$
$$= \sqrt{25 + 4} = \sqrt{29} = 5.385$$

The procedure for computing the slope of a line and the slant line distance of a line segment is as follows:

1. Sketch the x- and y-axes.
2. Locate the two points.
3. Calculate the rise (Δy) and the run (Δx). Do the results agree with the sketched distances?
4. Compute the slope:

$$\text{Slope} = \frac{\text{rise}}{\text{run}} = \frac{\Delta y}{\Delta x}$$

The slope is:

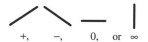

$+, \quad -, \quad 0, \quad \text{or} \quad \infty$

5. Determine the slant line distance (Δs):

$$\Delta s = \sqrt{(\Delta x)^2 + (\Delta y)^2}$$

For a three-dimensional line, such as in the following figure, the slant line distance (Δs) is

$$\Delta s = \sqrt{(\Delta x)^2 + (\Delta y)^2 + (\Delta z)^2}$$
$$= \sqrt{(3)^2 + (5)^2 + (7)^2}$$
$$= \sqrt{9 + 25 + 49}$$
$$= 9.110$$

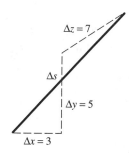

Time is used to keep track of the movement of objects. There are two ratios, or **rates,** that involve movement and time. These (time-related) rates are **velocity** and **acceleration.** Distance, velocity, and acceleration are computed using simple formulas.

If the velocity is constant, then

$$\Delta s = v \, \Delta t \quad \text{where } v \text{ is a constant}$$

Example If $v = 5$ m/s east and $\Delta t = 3$ s, then

$$\Delta s = 5 \text{ m/s E} \times 3 \text{ s} = 15 \text{ m E}$$

If the acceleration (a) is constant, then

$$s = \frac{1}{2}at^2$$
$$v = at$$
$$a = \text{constant}$$

Example   If $a = 6$ m/s^2 north and $t = 4$ s, then

$$s = \frac{1}{2} \times 6 \text{ m/s}^2 \text{ N} \times (4 \text{ s})^2 = 48 \text{ m N}$$
$$v = 6 \text{ m/s}^2 \text{ N} \times 4 \text{ s} = 24 \text{ m/s N}$$

Glossary of Math Words

Acceleration (ak′-sel-ur-ay′-shun)   A measure that describes a change in velocity or speed when related to time; expressed as $\Delta v / \Delta t$.

Angle (an′-gl)   The shape made by two straight lines that meet at a point.

Cartesian coordinate (car-tee′-see-an ko-or′-din-at) **system**   A system that describes the location of points on a two-dimensional, flat surface. It consists of a horizontal line (x-axis) intersected by a vertical line (y-axis). The intersection is known as the **origin;** the four corners of the intersection are known as **right angles.**

Deceleration (dee-sel-ur-ay′-shun)   The negative, or opposite, of acceleration.

Delta Δ (del′-ta)   The Greek letter used in math to indicate a change or difference.

Depth (depth′)   The name given to the measure of distance that is **into** or **out from** the two-dimensional Cartesian coordinate system; the third dimension.

Dimension, fourth (die-men′-shun, forth′)   The dimension believed to be **time.**

Distance (dis′-tans)   A one-dimensional measure along either horizontal, vertical, or slant line; when distance is expressed as a scalar quantity, the direction of the line is ignored.

Grade (grayde′) **of a road**   The measure of a road's slope; it is the value of the slope multiplied by 100. The value of a grade is expressed as a **percent.**

Horizontal (hor-ih-zon′-tul) **distance**   The distance measured along the horizontal real number line.

Hypotenuse (hy-pot′-uh-noos)   The slant line side of a right triangle; also the longest side.

Initial value (ih-ni′-shul val′-yoo)   The value of a measure when **time** is defined as **zero.**

Point (poynt′) **location**   In the two-dimensional Cartesian coordinate system, the horizontal distance (x) from the origin, followed by the vertical distance (y) from the origin; it is written (x, y).

Rate, time-related (rayte′)   A slope whose horizontal distance is **time;** thus, time is the denominator unit of this ratio.

Right triangle (ryte′ try′-an-gl)   A triangle that has one horizontal side, one vertical side, and a third (slant line) side known as the **hypotenuse.**

Rise (ryze′)   The vertical side of a right triangle; a vertical change in distance that is often written Δy.

Run (run′)   The horizontal side of a right triangle; a horizontal change in distance that is often written Δx.

Slant line (slant lyne′)   A line that is neither horizontal nor vertical; it describes the hypotenuse of a right triangle.

Slope (slowpe′)   The ratio of a change in vertical distance (**rise**) to a change in horizontal distance (**run**), often expressed as $\Delta y/\Delta x$, or rise/run. A line may have one of four types of slope:

Positive	Zero, or horizontal
Negative	Undefined, or vertical

Speed (speed′)   A measure that describes a change in distance with respect to time, ignoring the direction of the change; speed is the absolute value of **velocity.**

Subscript (sub′-script)   A number or letter written in the lower right corner of another number or letter; for the term s_0, the **0** is the subscript.

Time (tyme′)   The dimension believed to be the fourth dimension; **time** is involved in the explanation of motion.

Triangle (try′-an′gl)   A two-dimensional, closed surface that consists of three straight-line sides and, therefore, three corners known as **angles.**

Velocity (vuh-los′-i-tee)   A measure that describes the directional change (vector) of a distance when related to time, expressed as $\Delta s/\Delta t$.

Vertical (ver′-ti-cl) **distance**   The distance measured along the vertical real number line.

Chapter 19 Test

Follow instructions carefully:

> *On a separate piece of paper, write the answers to the following questions. Do* not *write on these pages.*

> *When you are finished, compare your answers with those given in Appendix B.*

> *Record the date, your test time, and your score on the chart at the end of this test.*

1. Determine the slope of a line constructed between points $A(3, 2)$ and $B(5, -4)$.

2. Determine the slope of a line constructed between points $A(3, 2)$ and $B(7, 2)$.

3. Determine the slope of a line constructed between points $A(3, 2)$ and $B(1, -6)$.

4. Determine the slope of a line constructed between points $A(3, 2)$ and $B(3, -5)$.

5. Determine the grade of a road that rises 3 ft over a horizontal distance of 80 ft.

6. Determine the grade of a road that falls 2 m over a horizontal distance of 80 m.

7. Determine the slant line distance between points $A(3, 2)$ and $B(5, -4)$.

8. Determine the slant line distance between points $A(3, 2)$ and $B(1, -6)$.

9. Compute the three-dimensional slant line distance where $\Delta x = 6.5$ cm, $\Delta y = 5.4$ cm, and $\Delta z = 8.1$ cm.

10. Compute the three-dimensional slant line distance where $\Delta x = 4$ ft, $\Delta y = -6$ ft, and $\Delta z = 2$ ft.

11. Compute the average velocity of a person who has required 4 hours to walk 14 miles to the west.

12. Compute the average velocity of a runner, in m/s, who runs 1800 m east in 15 min.

A car starts at a reference point and travels north at a constant velocity of 45 mi/hr for 4 hr.

13. How far has the car traveled in 4 hours?

14. Construct a graph of distance traveled versus time.

15. Construct a graph of velocity versus time.

A woman walks a distance of 7 miles east from her home in 2 hours.

16. Compute her average velocity.

17. Construct a graph of distance walked versus time.

18. Construct a graph of velocity versus time.

An aircraft travels south along a runway and accelerates from a rest position at a constant rate of 8 m/s^2.

19. Compute the aircraft's velocity after ten seconds of constant acceleration along the runway.

20. Compute the aircraft's distance traveled along the runway after ten seconds of constant acceleration.

Chapter 19 Test Record

DATE	TIME	SCORE

Refer to

- Appendix B for the correct answers to this test.
- Appendix C if taking this test required too much effort.
- Appendix D for **Memory Methods** assistance.
- Appendix E if your test scores are decreasing.

When a problem seems difficult, find one like it in this chapter. Then study that (and the related material) again.

Develop additional 3 × 5 cards for those ideas, problems, and procedures that caused you difficulty.

Part 6 Review Test

CHAPTER 18
Averages and Percents

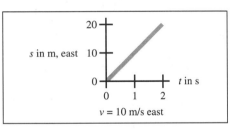

$$\frac{4+8}{20} = 0.60$$

$$\text{or } 60\%$$

CHAPTER 19
Slopes and Rates

Instructions

Review your 3 × 5 cards before you take this test.

Complete this test.

You may find some ideas, problems, or procedures difficult or confusing. Develop 3 × 5 cards for them.

Part 6 Review Test

1. Five families consist of 4, 9, 2, 7, and 3 people. Determine the mean average number of people in these five families.

2. Determine the location of the median (average) for Problem 1.

3. Determine the percent of red buttons if there are 16 red ones in a box of 128.

4. In Problem 3, there are also 30 green, 28 blue, and 54 black buttons. Determine the percent of green, blue, and black buttons. Construct a bar graph, labeling the horizontal axis with the 4 colors and the vertical axis **"Percent."** Display the percent values on this bar graph.

5. An average of $12/week is deposited for 8 weeks. Determine the deposit total.

A school consists of 12 classes of the following numbers of students:

$$4, \quad 16, \quad 10, \quad 9, \quad 63, \quad 9, \quad 18, \quad 10, \quad 12, \quad 16, \quad 6, \quad 10$$

6. Determine the (mean) average class size.

7. Determine the median average class size.

8. Determine the mode averages for these classes. Display the results on a bar graph.

9. Determine what percent of all the classes have ten students.

10. Joan deposits $15/week for four weeks and $21/week for eight weeks. Determine her average weekly deposit.

11. Determine the slope of a line constructed between points $A(-2, -4)$ and $B(3, 5)$. Display this line on a graph, and label its endpoints.

12. Determine the slope of a line constructed between points $A(-3, 6)$ and $B(7, -1)$. Display this line on a graph, and label its endpoints.

13. Determine the slant line distance between points $A(-3, 4)$ and $B(5, 10)$. Display this line on a graph, and label its endpoints.

14. Determine the slope of a line constructed between points $A(-2, -6)$ and $B(5, -4)$. Display this line on a graph, and label its endpoints.

15. Determine the slant line distance between points A and B in Problem 14.

16. Determine the slope of a line constructed between points $A(-2, 7)$ and $B(5, -4)$. Display this line on a graph, and label its endpoints.

A car travels at a constant velocity of 45 mi/hr to the south.

17. At the end of one hour, how far has the car traveled?

18. At the end of 3 hours, how far has the car traveled?

19. Using the following figure, sketch the graph of distance traveled (y) versus time (x) for the same three hours.

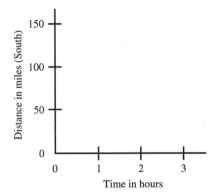

20. Using the following figure, sketch the graph of velocity (y) versus time (x) for the three hours.

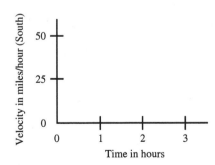

Part 6 Review Test Record

DATE	TIME	SCORE

Refer to

- Appendix B for the correct answers to this test.
- Appendix C if taking this test required too much effort.
- Appendix D for **Memory Methods** assistance.
- Appendix E if your test scores are decreasing.

When a problem seems difficult, find one like it near the page referenced with the answers in Appendix B. Study the related material.

Develop additional 3×5 cards for those ideas, problems, and procedures that caused you difficulty.

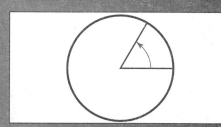

$1, 3, 5, 7, \ldots$

$1, 3, 5, 7, \ldots$

$1, 10, 100, 1000, \ldots$

1, 10, 100, 1000, . . .

Sequences and Series

Number Sequences

In our daily world it is useful to examine the past so that we can attempt to predict the future. We study information (history) so that we can estimate, or forecast, the condition of a business, the weather, the nation's economy, or the world's supply of raw materials and natural resources at a later time. Therefore, we must learn how to organize data, and best display our results, so that others can easily understand what we are trying to tell them.

When we examine a collection (or list) of numbers, we can **sequence** them so that they become more meaningful to us. When used by mathematicians, the words **numerical sequence** mean "a collection of numbers arranged in a special order." For example, the numbers

$$9, \quad 3, \quad 15, \quad 6, \quad 12$$

when arranged in an increasing numerical sequence become

$$3, \quad 6, \quad 9, \quad 12, \quad 15$$

Can you **predict** what the next number would be, now that these numbers have been arranged in numerical sequence? The next number would be **18.** Why? We will examine the reason later in this chapter.

Each number in a sequence is known as a **term.** The sequence 3, 6, 9, 12, 15 contains five terms:

- The first term in this sequence is **3.**
- The fifth term in this sequence is **15.**

A sequence that has a last (final) term is known as a **finite sequence.** A finite sequence has a de*finite* start and a de*finite* end. The above sequence is a **five-term, finite sequence.** It is also known as a **progression** because it progresses from smaller numbers to larger numbers along the number line. (A progression can also progress from larger to smaller numbers.)

A sequence that is believed to continue forever, and therefore has no final term, is known as an **infinite sequence** of terms. Three side-by-side dots (. . .) are used to indicate that a sequence continues forever.

Study the following infinite sequence:

$$2, \quad 4, \quad 6, \quad 8, \ldots$$

The three side-by-side dots indicate that this sequence has no end. These dots together are known as an **ellipsis.**

There are situations where we wish to add together all the terms in a sequence. The **vector sum** is computed. The vector sum of a sequence is also known as the **series** of that sequence.

A sequence is a list of numbers. A series is the operation of combining (adding or subtracting) these numbers.

Study the following examples.

Example A five-term sequence consists of

$$3, \quad 6, \quad 9, \quad 12, \quad 15$$

  a. Determine its series.
  b. Compute the value (sum) of this series.

Answer

  a. The series is

$$3 + 6 + 9 + 12 + 15$$

  b. The value (Σ) of this series is

$$\Sigma = 3 + 6 + 9 + 12 + 15$$
$$= 45$$

Note in the previous example that Σ is the Greek letter sigma and is the math symbol that indicates a **sum,** as indicated in Chapter 18. The **summation symbol** is further illustrated in the following example.

Example Johnny starts a bank account with $64. He then withdraws $32, deposits $16, withdraws $8, deposits $4, withdraws $2, and deposits $1.

  a. Write this bank account activity as a sequence.
  b. Compute the value of the series for this sequence.

Answer

  a. $64, $-$32, $16, $-$8, $4, $-$2, $1
     (This is a seven-term, finite sequence.)
  b. $\Sigma = $64 - $32 + $16 - $8 + $4 - $2 + 1
     $= 43

The $43 is the value of the series for this sequence. It represents the amount of money remaining in Johnny's bank account.

Exercise Set 20.1

The answers to the exercises in this chapter are given in Appendix A.

Work the following exercises.

1. A sequence consists of eight terms. The first four terms are 2, 4, 6, and 8.

   a. Write the eight terms in this sequence.

   b. The series for this sequence is

   c. The value of the series is

2. A sequence consists of seven terms. The first four terms are 1, −3, 5, and −7.

   a. Write the seven terms in this sequence.

   b. The series for this sequence is

   c. The value of the series is

3. Jane is given a jar containing three marbles. She is instructed to double the number of marbles in the jar every week.

    a. At the end of five weeks, how many marbles are in the jar?

    b. The (six-term) sequence that represents the number of marbles in the jar each week is

    c. The series for this sequence is

    d. The number of marbles in the jar at the end of five weeks is

4. A bush is known to look healthier if it is **cut back** (trimmed) to one-half of its height each year. During the remainder of the year, the bush grows to three times its trimmed height. At the end of the first year of growth, the bush is 16 in. tall before trimming.

    a. What is the height of the bush before trimming at the end of the fourth year?

    b. Write the seven terms in this sequence.

    c. The value of the seventh term is

        (This is the height of the bush before trimming at the end of the fourth year.)

5. A bank account starts with $50 and has added to it

    $20 the first month,
    $40 the second month,
    $50 the third month,
    $30 the fourth month,
    $60 the fifth month, and
    $10 the sixth month.

    a. What is the amount of money in the bank account at the end of the third month?

    b. Write the seven terms for this bank account sequence.

    c. The value of the series is

        (This is the amount of money in the bank account at the end of the sixth month.)

Sigma Notation for Sequences and Series

Mathematicians have developed a notation using the Greek **summation symbol sigma** (Σ) that tells us both the sequence terms and its series. In this **sigma notation,** the **lower limit** (first term) is indicated at the bottom of the sigma symbol. The upper limit (last term) is indicated at the top of the sigma symbol. The letter n is used as an **index;** n is always an integer. It tells us how to write each term in the series of the sequence. For example:

$$\sum_{n=1}^{n=5} 2^n \quad \text{where } 2^n \text{ is the \textbf{form} of each term}$$

with arrows labeling the upper limit ($n=5$) and lower limit ($n=1$).

The value of the index (n) ranges from the **lower limit** of 1 to the **upper limit** of 5. Note that the value of n increases in integer steps: 1, 2, 3, 4, 5. Therefore, there are five terms in the series.

The series begins with the first term, where $n = 1$, and ends with the last term, where $n = 5$. Each term is **formed** from 2^n.

Study the following example.

⬤Example

$$\sum_{n=1}^{n=5} 2^n = 2^1 + 2^2 + 2^3 + 2^4 + 2^5$$
$$= 2 + 4 + 8 + 16 + 32$$
$$= 62$$

The sequence for this series is 2, 4, 8, 16, 32. The value of this five-term series is 62.

We can make numbers alternate in sign by using the sigma-and-index notation. Study the following example.

⬤Example

$$\sum_{n=0}^{n=3} (-1)^n \times (2n)$$

$$= (-1)^0(2 \times 0) + (-1)^1(2 \times 1) + (-1)^2(2 \times 2) + (-1)^3(2 \times 3)$$
$$= (1) \times (0) + (-1) \times (2) + (+1) \times (4) + (-1) \times (6)$$
$$= 0 - 2 + 4 - 6$$
$$= -4$$

The sequence for this series is 0, −2, 4, −6. The value of this four-term series is −4.

Note: In programming, the letters i, j, k, l, m, and n are often used to indicate integer steps.

Exercise Set 20.2

Work the following exercises.

1. Evaluate (determine the value of) the following four-term series:

$$\sum_{n=0}^{n=3} \frac{1}{n+1}$$

2. Evaluate the following series:

$$\sum_{n=1}^{n=5} (2n+1)$$

3. Evaluate the following series:

$$\sum_{n=0}^{n=4} (-1)^n 2^n$$

4. Evaluate the following series:

$$\sum_{n=1}^{n=4} 3^{n+1}$$

5. Evaluate the following series:

$$\sum_{n=1}^{n=3} \frac{n + 1}{2n + 1}$$

6. Evaluate the following series:

$$\sum_{n=-2}^{n=3} 2^n$$

MATH AT WORK

Arithmetic Progressions

A sequence that occurs often, both in math and in business, is the **arithmetic progression.** This sequence has the same numerical value **(distance)** from one term to the next term. This distance is known as the **common difference (d)** of an arithmetic progression.

Examine the arithmetic progression 3, 7, 11, 15. The common difference (d) can be computed by subtracting the value of one term from the value of the next term:

$$d = 15 - 11 = 4$$
$$d = 11 - 7 = 4$$
$$d = 7 - 3 = 4$$

Therefore, **4** is the common difference for this four-term arithmetic progression.

The series (S_4) for this four-term sequence is determined by combining the four terms:

$$S_4 = 3 + 7 + 11 + 15$$
$$= 36$$

This method for computing the series for an arithmetic sequence is the same as the method described previously in this chapter for computing the value of any series.

Note that the S_4 (spoken "*s* sub-four" where the 4 is referred to as the subscript to *S*) notation has the same meaning as the sigma (Σ) notation, summing from one to four:

$$\sum_{n=1}^{n=4}$$

Some arithmetic progressions contain a large number of terms. Also, the common difference (*d*) might be a fraction. Formulas are available that permit you to quickly determine the following:

- the value of a term,
- the value of the series, and
- the common difference.

Each term of a sequence is identified using the letter *a* and an integer subscript *n*:

a_1 is the first term, a_2 is the second term, a_3 is the third term, and a_n is the general term.

The value chosen for *n* identifies that term and the finite length of the sequence.

For the four-term sequence 3, 7, 11, 15, the terms are identified as follows:

$$a_1 = 3 \quad \text{(the first term)}$$
$$a_2 = 7 \quad \text{(the second term)}$$
$$a_3 = 11 \quad \text{(the third term)}$$
$$a_4 = 15 \quad \text{(the fourth term)}$$

The index letter *n* is used again. When a formula is being developed, the **general term** is written a_n and is spoken "*a* **sub-*n*.**" As each of the four terms is calculated, one at a time, the index letter n becomes first **1**, then **2**, then **3**, and then **4**.

There are two fundamental formulas that apply to the arithmetic progression. When the common difference (d) is known, and the value of the first term (a_1) is also known, then any other term (the nth term a_n) may be calculated using

$$a_n = a_1 + (n - 1)d$$

For a finite arithmetic progression, the value of the series that consists of n terms (S_n) may be computed using

$$S_n = \frac{n}{2}(a_1 + a_n)$$

Note that both the first term (a_1) and the nth term (a_n) must be known before S_n can be computed.

Study the following examples.

Example   The first term (a_1) of a five-term arithmetic progression is **7**, and the common difference (d) is **3.**

    a. Determine the other four terms in this sequence.

    b. Determine the value of this series.

Answer

    a. For $n = 1$, $a_1 = 7$ (given)
        For $n = 2$, $a_2 = 7 + (2 - 1) \times 3 = 7 + 3 = 10$
        For $n = 3$, $a_3 = 7 + (3 - 1) \times 3 = 7 + 6 = 13$
        For $n = 4$, $a_4 = 7 + (4 - 1) \times 3 = 7 + 9 = 16$
        For $n = 5$, $a_5 = 7 + (5 - 1) \times 3 = 7 + 12 = 19$
        Then

$$S_5 = \frac{5}{2}(7 + 19)$$

$$= \frac{5}{2}(26)$$

$$= 65$$

    b. This agrees with combining all five terms in the sequence:

$$S_5 = 7 + 10 + 13 + 16 + 19 = 65$$

Note: The term a_3 is considered to be the **arithmetic mean** of the series because it is the center of a mean average of the progression.

Example For a four-term arithmetic progression, the first term is **5** and the second term is **7.**

    a. Determine the other two terms in this sequence.

    b. Determine the value of the series.

Answer

    a. For $n = 1$, $a_1 = 5$ (given) $\Big\}$ Therefore, the common
        For $n = 2$, $a_2 = 7$ (given) difference (d) is **2.**
        For $n = 3$, $a_3 = 5 + (3 - 1) \times 2 = 5 + 4 = 9$
        For $n = 4$, $a_4 = 5 + (4 - 1) \times 2 = 5 + 6 = 11$
        Then

$$S_4 = \frac{4}{2}(5 + 11)$$

$$= \frac{4}{2}(16)$$

$$= 32$$

    b. This agrees with combining all four terms in the sequence:

$$S_4 = 5 + 7 + 9 + 11 = 32$$

When the terms of an arithmetic progression are displayed on the real number line, the distance between adjacent (side-by-side) terms is the same. This spacing (distance) is the common difference (d).

Study the following examples.

Example Indicate the following five terms of an arithmetic progression on the real number line:

$$7, \quad 10, \quad 13, \quad 16, \quad 19$$

Answer

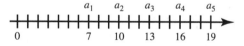

Example Indicate the following four terms of an arithmetic progression on the real number line:

$$5, \quad 7, \quad 9, \quad 11$$

Answer

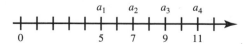

Example For a four-term arithmetic progression, the first term is **7** and the second term is **4.**

    a. Determine the other two terms.

    b. Determine the value of the series.

    c. Display the values on a real number line.

Answer

a. For $n = 1$, $a_1 = 7$ (given) $\left.\vphantom{\begin{matrix}a\\a\end{matrix}}\right\}$ Therefore, the value of
   For $n = 2$, $a_2 = 4$ (given) $\quad d$ is $(4 - 7)$ or $-3.$
   For $n = 3$, $a_3 = 7 + (3 - 1)(-3) = 7 - 6 = 1$
   For $n = 4$, $a_4 = 7 + (4 - 1)(-3) = 7 - 9 = -2$

b. Then

$$S_4 = \frac{4}{2}\,(7 + (-2)) = \frac{4}{2}\,(5) = 10$$

c. These values are displayed on the real number line as follows:

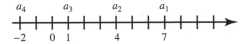

Note again that the word **progression** is also applied to sequences that progress from larger numbers to smaller numbers.

Exercise Set 20.3

Work the following exercises.

1. For a six-term arithmetic progression, the first term is **2** and the common difference is **3.**

   a. Determine the other five terms.

   b. Determine the value of the series for this sequence.

   c. Indicate the six terms of this sequence on the real number line.

2. For a five-term arithmetic progression, the first term is **3** and the common difference is **1.5.**

   a. Determine the other four terms.

   b. Determine the value of the series for this sequence.

   c. Indicate the five terms of this sequence on the real number line.

3. For a five-term arithmetic progression, the first term is **11** and the common difference is **−4.**

    a. Determine the other four terms.

    b. Determine the value of the series for this sequence.

    c. Indicate the five terms of this sequence on the real number line.

4. For a five-term arithmetic progression, the first term is **−5** and the second term is **−1.**

    a. Determine the other three terms.

    b. Determine the value of the series for this sequence.

    c. Indicate the five terms of this sequence on the real number line.

Geometric Progressions

A sequence that occurs often in math, business, and nature is the **geometric progression.** This sequence has the same numerical ratio from one term to the next term. This ratio is known as the **common ratio (r)** of a geometric progression.

Examine the four-term geometric progression 3, 6, 12, 24. We can compute the common ratio *(r)* by dividing the value of any term by the value of the previous term. Thus,

$$r = \frac{24}{12} = 2$$

$$r = \frac{12}{6} = 2$$

$$r = \frac{6}{3} = 2$$

Therefore, **2** is the common ratio for this four-term geometric progression.

The series for this four-term sequence is determined by combining the four terms:

$$S_4 = 3 + 6 + 12 + 24$$
$$= 45$$

This method for computing the series of a geometric progression is the same as the method described previously in this chapter for computing the value of any series.

The index letter *n* is used again. Its use in one of the geometric progression formulas results in an exponent (power) whose math notation is new to you. The term

$$r^{n-1}$$

is spoken "***r* to the *n*-minus-one power.**"

There are two fundamental formulas that apply to a geometric progression. When the common ratio *(r)* is known, and the value of the first term (a_1) is known also, then any other term (a_n) may be calculated using

$$a_n = a_1 r^{n-1}$$

For a finite geometric progression, the value of the series consisting of *n* terms (S_n) may be computed using

$$S_n = \frac{a_1 - ra_n}{1 - r}$$

Note that the first term (a_1), the *n*th term (a_n), and *r* must be known before S_n can be computed.

Also note that the common ratio *r* cannot be **1** because the denominator of S_n would become zero, which results in a meaningless answer. Therefore, a sequence of the same number (such as 3, 3, 3, 3, 3) which is a sequence with a common ratio of **1** is not a geometric progression!

The following figure illustrates how changing a picture from a display on a geometric scale to a display on a linear scale causes it to become distorted:

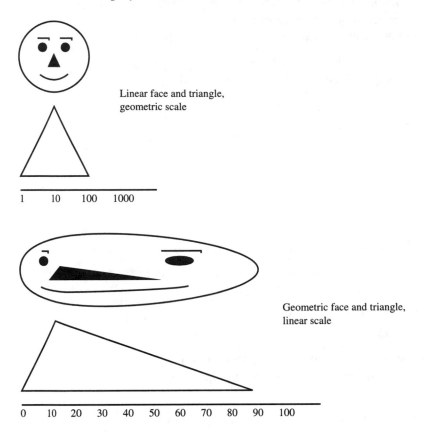

Linear face and triangle, geometric scale

Geometric face and triangle, linear scale

Study the examples on the following page.

MATH AT WORK

Example The first term of a four-term geometric progression is **2** and the common ratio is **3.**

    a. Determine the other three terms in this sequence.

    b. Determine the value of its series.

Answer

    a. For $n = 1$, $a_1 = 2$ (given)
       For $n = 2$, $a_2 = 2 \times 3^{2-1} = 2 \times 3^1 = 6$
       For $n = 3$, $a_3 = 2 \times 3^{3-1} = 2 \times 3^2 = 18$
       For $n = 4$, $a_4 = 2 \times 3^{4-1} = 2 \times 3^3 = 54$

    b. Then

$$S_4 = \frac{2 - 3 \times 54}{1 - 3}$$

$$= \frac{2 - 162}{-2}$$

$$= \frac{-160}{-2}$$

$$= +80$$

Note that this value for S_4 agrees with that obtained by combining the four terms in the sequence:

$$S_4 = 2 + 6 + 18 + 54 = 80$$

Example The first term of a five-term geometric progression is **7** and the second term is **21.** Determine the other three terms in this sequence.

Answer

For $n = 1$, $a_1 = 7$ (given)

Thus, $r = \dfrac{21}{7} = 3$

For $n = 2$, $a_2 = 21$ (given)
For $n = 3$, $a_3 = 7 \times 3^{3-1} = 7 \times 3^2 = 63$
For $n = 4$, $a_4 = 7 \times 3^{4-1} = 7 \times 3^3 = 189$
For $n = 5$, $a_5 = 7 \times 3^{5-1} = 7 \times 3^4 = 567$

The value of the series could be computed if desired. (Its value is 847.)

Note: The term a_3 is considered the **geometric mean** of the series because it is the geometric center of the progression.

Exercise Set 20.4

Work the following exercises.

1. The first term of a three-term geometric progression is **5** and the common ratio is **2.**

    a. Determine the other two terms in this sequence.

    b. Determine the value of its series.

2. The first term of a four-term geometric progression is **5** and the second term is **20.**

    a. Determine the other two terms in this sequence.

    b. Determine the value of its series.

The terms of a geometric progression are often displayed on a modified real number line. The real number line is normally constructed using the same (linear) spacing between numbers. This line may be modified so that each mark represents the exponent of a chosen **base.**

The most common base for this modified real number line is the **base 10.** The resulting markings are known as **logarithmic** or **log scale** markings. The following figure illustrates this modified real number line:

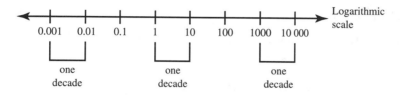

Any two consecutive markings shown on this log scale are ratios of ten to one. Therefore, each space is known as one **decade,** a word that was derived from Latin and that means **"a group of ten."** For example, a total of seven decades are displayed on the following log scale, real number line:

Psychologists have discovered that our five senses and our minds can process ratios better than differences whenever very large or very small numbers are involved. Therefore, you will see more frequent use of logarithmic scales as technological advances occur.

Note that on our log scale in the first figure above, the ratio from one end (10^4) to the other (10^{-3}) is

$$\frac{10^4}{10^{-3}} = 10^{4-(-3)} = 10^{4+3} = 10^7$$

This is a ratio of ten million to one! We can display numbers starting at 1×10^{-3} (0.001) and ending at 1×10^4 (10 000) using these seven decades of a log scale.

Next we shall examine more closely the markings on a single decade. We shall compare the spacing of its markings with the markings from **0** to **1** on the companion linear scale:

Note that as you proceed from left to right, the log scale, whole number markings become closer together.

Now compare three adjacent decades. (For marking simplicity, we shall use only numbers that are base 10 multiples of 1, 2, and 5.) The following is a three-decade log scale, shown with a companion linear scale:

Let's examine the spacing differences and ratios between the 2's and 1's.

- For the left decade (0.1 to 1):

  One spacing difference is $0.2 - 0.1 = 0.1$.
  The ratio is $0.2/0.1 = 2$.

- For the center decade (1 to 10):

  One spacing difference is $2 - 1 = 1$.
  The ratio is $2/1 = 2$.

- For the right decade (10 to 100):

  One spacing difference is $20 - 10 = 10$.
  The ratio is $20/10 = 2$.

The differences are not the same; the ratios are identical (**2**). We work with a log scale when we want to compare ratios.

Exercise Set 20.5

Work the following exercises.

1. On the following scale, examine the numbers that contain **5** and **1.** List their spacings (differences) and their ratios for each of the three decades.

2. On the following scale, examine the numbers that contain **1, 2,** and **5.**

a. List their spacings (differences) and their ratios for each of the three decades.

b. Compare these values with the given linear scale.

Study the following examples.

Example For a five-term geometric progression, the first term is **1** and the common ratio is **3.**

 a. Determine the other four terms.

 b. Determine the value of the series for this sequence.

 c. Display the five terms of this sequence on a log scale, real number line.

Answer

 a. For $n = 1$, $a_1 = 1$ (given)
 For $n = 2$, $a_2 = 1 \times 3^{2-1} = 1 \times 3 = 3$
 For $n = 3$, $a_3 = 1 \times 3^{3-1} = 1 \times 9 = 9$
 For $n = 4$, $a_4 = 1 \times 3^{4-1} = 1 \times 27 = 27$
 For $n = 5$, $a_5 = 1 \times 3^{5-1} = 1 \times 81 = 81$

 b. Then

$$S_5 = \frac{1 - 3 \times 81}{1 - 3}$$

$$= \frac{1 - 243}{-2}$$

$$= \frac{-242}{-2}$$

$$= +121$$

 c. The five terms are displayed as follows on the log scale, real number line:

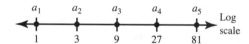

Take a small piece of paper and place it against the logarithmic scale. Make two marks on the paper: one at the location of a_1, the other at the location of a_2. Move this paper so that the first mark is aligned with a_2. The second mark should be aligned with a_3. Repeat this process for both a_4 and a_5, noting the equal spacing.

Example For a six-term geometric progression, the first term is **5** and the second term is **10**.

    a. Determine the other four terms.

    b. Determine the value of the series for this sequence.

    c. Display the six terms of this sequence on a log scale, real number line.

Answer

    a. For $n = 1$, $a_1 = 5$ The common ratio is $\dfrac{10}{5} = 2$.

        For $n = 2$, $a_2 = 10$

        For $n = 3$, $a_3 = 5 \times 2^{3-1} = 5 \times 4 = 20$

        For $n = 4$, $a_4 = 5 \times 2^{4-1} = 5 \times 8 = 40$

        For $n = 5$, $a_5 = 5 \times 2^{5-1} = 5 \times 16 = 80$

        For $n = 6$, $a_6 = 5 \times 2^{6-1} = 5 \times 32 = 160$

    b. Then

$$S_6 = \frac{5 - 2 \times 160}{1 - 2}$$

$$= \frac{5 - 320}{-1}$$

$$= \frac{-315}{-1}$$

$$= +315$$

    c. The six terms are displayed as follows on a log scale, real number line:

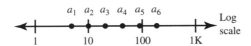

Note the unequal spacing of terms on a linear scale. The 2-to-1 spacing is more apparent, however:

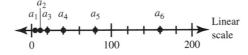

Exercise Set 20.6

Work the following exercises.

1. For a five-term geometric progression, the first term is **4** and the common ratio is **5.**

    a. Determine the other four terms.

    b. Determine the value of the series for this sequence.

    c. Display the five terms of this sequence on the following log scale, real number line:

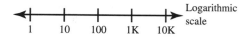

2. For a six-term geometric progression, the first term is **3** and the second term is **30.**

    a. Determine the other four terms.

    b. Determine the value of the series for this sequence.

    c. Display the six terms of this sequence on the following log scale, real number line:

3. For a five-term geometric progression, the first term is **2** and the second term is **20.**

    a. Determine the other three terms.

    b. Determine the value of the series for this sequence.

    c. Display the five terms of this sequence on the following log scale, real number line:

Logarithmic scale

1   10   100   1K   10K   100K   1M

4. The musical note **middle C** has a frequency of 523 cycles per second, known as 523 hertz, or 523 Hz. All other **C's** on the musical scale are spaced **one octave** apart, which is a common ratio of 2 to 1. Determine the frequencies for the following table of C's:

Musical Note	Frequency
Two octaves below	
One octave below	
Middle C	
One octave above	
Two octaves above	
Three octaves above	
Four octaves above	

Statistics Applications

The study of numbers causes many of us to become more curious regarding their meaning and usefulness. For example:

> A child can be shown the value of saving or investing money if someone demonstrates the growth of a $10 investment during a span of several years. That child is usually impressed.

> Someday we may become responsible for planning the construction of office space for a business. It then becomes necessary to examine the growth of that business, including how that growth will affect the need for office personnel.

> The use of our natural resources is a concern to many of us today. We may wish to estimate how soon a resource such as oil will be exhausted. Or we may wish to forecast how much of a scarce metal is required if many vehicles will be battery-powered! We may wish to estimate how much of a renewable resource, such as solar energy, can be efficiently converted to a form that we can use wisely.

> Others of us may be concerned that a city will become overpopulated. We will have to estimate how rapidly the city will grow, how much land will be available, and how dense the population can become if we are to **live and breathe.**

All of these topics require that we either know, or can estimate, the type of growth involved. Is the change arithmetic or geometric, or does some other pattern exist? Specialized courses and books include formulas and charts used to more quickly and accurately perform these estimates.

The following simplified examples were prepared to assist you in learning to apply your new math skills without either special courses or special books. Note that these simplified problems provide some, but not necessarily all, of the information useful in making well-informed decisions. Other important considerations, such as the short-term versus long-term benefits and costs of different courses of action, are beyond the scope of this book.

Study the examples on the following pages.

Example   A savings account is started with a $10 deposit. Each year 5% of the money in the account is added as **interest.** Assume that no other money is deposited or withdrawn. How many years are required for the bank account to either become or slightly exceed $20?

Answer   We shall prepare a chart so that we may (1) develop a record of the money in the bank account at the beginning of each year. Then we can (2) compute and note the amount of interest credited to the account at the end of that year. The chart follows. If you have a calculator that has a storage key (STO), then practice storing the account balance so that you can recall (RCL) that balance when you wish to add the interest earned to the nearest cent.

Number of Years	Account Balance	Interest Earned
1	$10.00	$0.50
2	10.50	0.52
3	11.02	0.55
4	11.57	0.58
5	12.15	0.61
6	12.76	0.64
7	13.40	0.67
8	14.07	0.70
9	14.77	0.74
10	15.51	0.78
11	16.29	0.81
12	17.10	0.86
13	17.96	0.90
14	18.86	0.94
15	19.80	0.99
16	20.79	—

When you study more advanced texts, you will discover that a formula has been developed to replace all of the computations in the preceding example. This formula will permit you, in one calculation, to determine the account balance for any desired year.

Example   A copper mine is believed to contain 240 000 tons of copper that can be easily extracted (mined). The first year it will be possible to extract 10 000 tons of copper. Each succeeding year, newly available equipment and more efficient production techniques will yield a 20% increase of copper extracted compared to the previous year. How many years will be required to extract all of the copper from the mine?

Answer   For this problem, we must compute, and then note, the number of years, the amount of copper extracted that year (annual yield), and the total amount of copper extracted. First, let's make the following chart:

Number of Years	Annual Yield (in Tons)	Total Copper Extracted (in Tons)
1	10 000	10 000
2	12 000	22 000
3	14 400	36 400
4	17 280	53 680
5	20 736	74 416
6	24 883	99 299
7	29 860	129 159
8	35 832	164 991
9	42 998	207 989
10	51 598	259 587

The mine should be **exhausted** within ten years. Therefore, the real estate agent for the land that contains the copper would request a ten-year lease.

Example   A building is being planned for a newly formed company. The architect estimates that 200 square feet (sq ft) of space per employee will be needed for offices, corridors, restrooms, and storage areas. The president of the company wants the building to have enough space for ten years of growth, at 10% per year, starting with an office staff of thirty persons. Determine the building size, in square feet. Also determine how many square feet the president can rent each year to other companies.

Answer   First, determine the number of offices the new company will need each year and the number of square feet needed by the newly formed company. Once the square feet needed in the tenth year is known (the size of the building that the president desires), then it will be possible to compute the number of square feet that can be rented each year. The answers are given in the following chart:

Number of Years	Number of Offices	Square Feet Needed	Square Feet Available to Rent
1	30	6 000	8000
2	33	6 600	7400
3	36	7 200	6800
4	40	8 000	6000
5	44	8 800	5200
6	48	9 600	4400
7	53	10 600	3400
8	58	11 600	2400
9	64	12 800	1200
10	70	14 000	—

Example The cost for constructing a 325 000-square-foot state building is $32 500 000, or $100 per square foot. The construction cost will be borrowed at the rate of 5% per year. Adding solar-heating and energy-conservation features to the building increases the construction cost by $3 100 000 (9.54%).

The savings from using solar energy will be approximately $300 000 per year when compared to the cost of heating, ventilating, and air conditioning (HVAC) using either gas or oil as the fuel. Can the $300 000 annual savings alone pay off the additional cost of the $3 100 000 plus its interest at 5% per year? If so, how many years will be required to pay off the added solar construction costs?

Answer For this problem, we must compute the solar energy investment cost each year beginning with the initial cost of $3 100 000. By adding the 5% loan interest to each year's solar investment cost, and then subtracting the $300 000 annual savings, we shall have the next year's solar investment costs. We repeat this formula until the solar energy investment cost becomes either zero or a negative number. At this point, we will know the number of years required to completely repay the solar energy investment cost. The following chart shows the calculations:

Number of Years	Solar Energy Investment Cost	Loan Interest at 5%	Annual Solar Energy Savings	Next Year's Solar Energy Investment Cost
0	$3 100 000.00	$155 000.00	$300 000.00	$2 955 000.00
1	2 955 000.00	147 750.00	300 000.00	2 802 750.00
2	2 802 750.00	140 137.50	300 000.00	2 642 887.50
3	2 642 887.50	132 144.38	300 000.00	2 475 031.90
4	2 475 031.90	123 751.59	300 000.00	2 298 783.50
5	2 298 783.50	114 939.17	300 000.00	2 113 722.60
6	2 113 722.60	105 686.13	300 000.00	1 919 408.80
7	1 919 408.80	95 970.44	300 000.00	1 715 379.20
8	1 715 379.20	85 768.96	300 000.00	1 501 148.20
9	1 501 148.20	75 057.41	300 000.00	1 276 205.60
10	1 276 205.60	63 810.28	300 000.00	1 040 015.90
11	1 040 015.90	52 000.79	300 000.00	792 016.65
12	792 016.65	39 600.83	300 000.00	531 617.49
13	531 617.49	26 580.87	300 000.00	258 198.36
14	258 198.36	12 909.92	300 000.00	−28 891.72
15	−28 891.72			

The $300 000 in annual energy savings will have entirely paid off the solar cost within 15 years. The state will continue to save $300 000 in energy costs for each year that it uses the building beyond this 15-year payoff period.

Exercise Set 20.7

Work the following exercises.

1. A savings account is started with a $10 deposit. Each year, 6% of the money in the account is added as interest. Assume that no other money is deposited or withdrawn. How many years are required for the bank account to equal or slightly exceed $20?

2. A town decides to construct a reservoir to hold enough water to supply the town's needs for at least two years (at any one time during the next ten years). The town planners have estimated that the town's water needs will grow at the rate of 14% per year for the next ten years. If the town is presently using 100 million gallons of water per year, then what should the capacity of the reservoir be, in millions of gallons?

3. A scientist working for an oil company estimates that a newly discovered oilfield contains 37 million barrels of oil. During the first year of oilfield operation, 6 million barrels of oil are extracted (pumped). Each year thereafter, 7% less oil can be extracted from this field. How long will the oilfield be able to supply oil, assuming that the original estimate is correct?

4. The cost of a house purchased today is $150 000. Inflation is causing the value of the house to **appreciate** (increase) at the rate of 7.8% per year. Estimate the value of the house six years from now, to the nearest dollar.

Summary

People exploring the arts and sciences often study events and activities by arranging them into orderly sequences. The math meaning of the word **sequence** is "a collection of numbers arranged in a special order." An example of a numerical sequence is

$$1, \quad 3, \quad 6, \quad 10, \quad 15, \quad 21$$

Each number in a sequence is known as a **term.** The first term (a_1) in this five-term sequence is **1;** the fifth term (a_5) is **21.**

A sequence that has a last, or final, term is known as a **finite sequence.** A sequence that goes on forever (...) is known as an **infinite sequence.** The sequence **1, 3, 5, 7, 9, ...** is an infinite sequence.

The vector sum, indicated by **sigma notation** (Σ), of the terms of a sequence is known as a **series.** The series for the sequence **2, 4, 6, 8, 10, 12** is

$$\Sigma = 2 + 4 + 6 + 8 + 10 + 12$$
$$= 42$$

The series of a sequence can be written in shortened form by using the index letter n; it is an integer that has, in the following example, a lower limit **(0)** and an upper limit **(4).** It is used to describe the form of each term **(3^n).**

Example

$$\sum_{n=0}^{n=4} 3^n = 3^0 + 3^1 + 3^2 + 3^3 + 3^4$$
$$= 1 + 3 + 9 + 27 + 81$$
$$= 121$$

A sequence whose adjacent (side-by-side) terms have a common difference (d) is known as an **arithmetic progression.** Each term (a_n) of an arithmetic progression can be computed using the formula

$$a_n = a_1 + (n - 1)d$$

For a finite arithmetic progression, the value of the series (S_n) can be computed using the formula

$$S_n = \frac{n}{2}(a_1 + a_n)$$

The terms of an arithmetic progression are equally spaced on the real number line. Therefore, the arithmetic progression **1, 4, 7, 10** is displayed as shown in the following figure:

A sequence whose adjacent terms have a **common ratio (r)** is known as a **geometric progression.** Each term (a_n) of a geometric progression can be computed using the formula

$$a_n = a_1 r^{n-1}$$

For a finite geometric progression, the value of the series (S_n) can be computed using the formula

$$S_n = \frac{a_1 - ra_n}{1 - r}$$

The terms of a geometric progression are equally spaced on the log scale, real number line. Therefore, the geometric progression **1, 2, 4, 8** is displayed as shown in the following figure:

We should be able to estimate various types of growth and decay of materials. Often the rate of growth and decay depends upon how much material has been used or upon how much material remains.

Complicated formulas have been developed to simplify the growth and decay computations. Logic and reasoning will also permit us to compute the desired values without using the complicated formulas. The results from using logic and reasoning will usually be as accurate as those obtained using the complicated formulas.

Glossary of Math Words

Arithmetic mean (air-ith-met′-ik meen′)   Any term of an arithmetic progression that is between the first term and the last term.

Arithmetic progression (air-ith-met′-ik pro-gresh′-un)   A sequence of numbers that has a common difference (d) between adjacent terms.

Common difference (com′-mun dif′-fur-ense)   The difference between adjacent terms of an arithmetic progression; symbolized by the letter d.

Common ratio (com′-mun ray′-she-oh)   The ratio between adjacent terms of a geometric progression; abbreviated r.

Decade (dek′-ayde)   A ratio of ten to one; a distance on the log scale, real number line that has a ratio of ten to one.

Finite sequence (fy′-nyte see′-kwense)   A sequence that has a last (ending) term.

Geometric mean (gee-o-met′-rik meen′)   Any term of a geometric progression that is between the first term and the last term.

Geometric progression (gee-o-met′-rik pro-gresh′-un)   A sequence of terms where the ratio between adjacent terms is the same.

Index (in′-deks)   The unknown integer in an expression (such as 3^n) that tells us how to write each term in the series of a sequence; the letter n, used in sigma notation, that indicates the integers to be used in each term of the sequence.

Infinite sequence (in′-fi-nit see′-kwense)   A sequence that continues forever and, therefore, has no ending; an infinite sequence is indicated by three dots (. . .).

Logarithm (log′-a-rith′-um)   The name given to the result of the inverse of an exponent; its abbreviation is **log.**

Log scale (log scayle′)   A real number line whose markings are spaced based on ratios of ten to one.

Numerical sequence (new-mair′-ih-kul see′-kwense)   Numbers arranged in a numerical order, usually from the smallest number to the largest number.

Octave (oc′-tiv)   In music, a spacing of eight sounds, or notes, that has a 2-to-1 frequency ratio between the identified notes.

Progression (pro-gresh′-un)   A group of terms that progresses from smaller numbers to larger numbers, or from larger numbers to smaller numbers; a progression is often displayed on the real number line.

Sequence (see′-kwense)   A collection of numbers arranged in a special order.

Series (seer′-eez)   The sum of the terms in a sequence; also known as the **vector sum.**

Sigma notation (sig′-ma no-tay′-shun)   The use of the Greek letter sigma (Σ) that includes an **index** identifying the upper limit and the lower limit of the sum of terms.

Subscript (sub′-skript)   A number or letter placed below and to the right of a base, such as the **4** in S_4.

Sum (sum′)   *See* Sigma notation.

Summation symbol Σ (sig′-ma)   The Greek summation symbol used to indicate that the sum of terms is to be performed.

Term (turm′)   The name given to each number in a sequence.

Vector sum (vek′-ter sum′)   *See* Series.

Chapter 20 Test

Follow instructions carefully:

On a separate piece of paper, write the answers to the following questions. Do not write on these pages.

When you are finished, compare your answers with those given in Appendix B.

Record the data, your test time, and your score on the chart at the end of this test.

1. For the sequence **1, 4, 7, 10, 13,** determine the next term.

2. For the sequence **5, 10, 20, 40,** determine the next term.

3. For the sequence **2, 5, 9, 14, 20,** determine the next term.

4. Is the sequence **5, 10, 20, 40** a finite or an infinite sequence?

5. Is the sequence **5, 10, 20, . . .** a finite or an infinite sequence?

6. What is the common difference for the progression **6, 12, 18, 24?**

7. What is the value of the series of Problem 6?

8. What is the common ratio for the progression **6, 36, 216?**

9. What is the value of the series of Problem 8?

10. Determine the third term of the following sequence:

$$\sum_{n=1}^{n=6} (2n + 1)$$

The musical note **middle A** has a frequency of 440 Hz. All other **A's** on the musical scale are spaced one **octave** apart. An **octave** is a spacing of eight notes that has a common frequency ratio of 2 to 1 between the identified notes.

11. Determine the frequency of the musical note **A** one octave below **middle A.**

12. Determine the frequency of **A** two octaves above **middle A.**

13. Determine the frequency of **A** three octaves below **middle A.**

A savings account is started with $10 deposit. Each year, 7% of the money in the account is added as **interest.** Assume that no other money is deposited or withdrawn.

14. How much money will be in the account at the end of three years, including interest?

15. How much money will be in the account at the end of five years, including interest?

16. How much interest has been paid at the end of four years?

17. How much interest has been paid at the end of seven years?

A child is born today. The parents start a savings account so that the child, at 18 years of age, will have enough money to attend a four-year college. Today a state college tuition is $4000 per year and is increasing at an annual rate of 4%.

18. Determine the amount of college tuition for the first year the child is to go to college.

19. Determine the amount of college tuition for the fourth year the child is to go to college.

20. Determine the total amount of college tuition the child will require for the entire four years of college.

Chapter 20 Test Record

DATE	TIME	SCORE

Refer to

- Appendix B for the correct answers to this test.
- Appendix C if taking this test required too much effort.
- Appendix D for **Memory Methods** assistance.
- Appendix E if your test scores are decreasing.

When a problem seems difficult, find one like it in this chapter. Then study that (and the related material) again.

Develop additional 3 × 5 cards for those ideas, problems, and procedures that caused you difficulty.

Circular Measure

Review of Lines and Curves

Recall that all smooth curves consist of

- straight lines, or line segments;
- circles, or portions of circles; or
- a combination of line segments or portions of circles.

This fact was discovered in approximately A.D. 1800 by the mathematician Jean Fourier. (See Chapter 17.)

In Chapter 19 we noted that in a two-dimensional Cartesian (x, y) coordinate system, one measure of change is the slope of a straight line. This straight line (linear) change is

$$\text{Linear change, or slope} = \frac{\Delta y}{\Delta x}$$

This linear change becomes a unitless ratio if both measures have the same units, such as millimeters divided by millimeters.

When we study linear change, we say that a moving body **generates** a straight line segment as it moves from one place (**A**) to another place (**B**). This body must always move in the same direction to generate a straight line segment. The line segment, whose length is $\Delta\mathbf{s}$, may be considered to be a vector, with a tail at **A** and a tip at **B** as shown in the following figure:

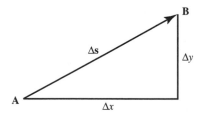

Recall from Chapter 19 that the **angle** between Δy and Δx is a right angle.

Radians: A Measure of Angular Distance

When we study circular motion, also known as *angular change,* we begin with a vector located on the real number line, the **reference axis.** The vector length remains a constant value **r,** as shown in the following figure:

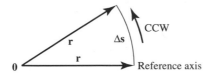

The tail of the vector is fastened at the origin **0.** The vector is then turned **counterclockwise (CCW),** as shown in the following figure:

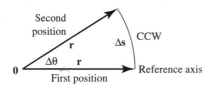

The vector tip generates a portion of the circumference of a circle. The distance along this portion of a circumference is **Δs,** where **Δs** is an arc. (An **arc** is the distance along a portion of the circumference of a circle.)

The origin **0** is the center of a circle whose radius is **r.**

Because the tail of the vector cannot change its position, the entire vector has turned an angular distance θ, pronounced "**theta.**" The **change** in angle, measured from the first position of the vector to its second position, is known as "delta theta" (Δθ):

This angular change (Δθ) is the ratio of two distances: the portion of an arc (**Δs**) divided by the length of a vector (**r**):

$$\Delta\theta = \frac{\Delta s}{r}$$

The units of this ratio, such as inches divided by inches, will cancel. Thus, angular change (Δθ) is a unitless ratio.

Angular change $\Delta\theta$ (also known as *angular distance* θ) can be a measure of turning motion. The **angle** through which a line (vector) turns does **not** depend upon the length of that line. For example, a line that is ten centimeters long may turn through the same **angular distance** (θ) as a line that is five centimeters long:

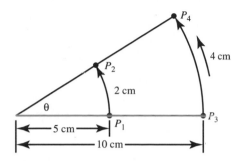

Note in this figure that the length of the circle arc that is generated **does** change; the distance from P_3 to P_4 is twice the distance from P_1 to P_2. Thus, angle θ remains constant because it is the same ratio value.

For the line that is 5 cm long:

$$\theta = \frac{2 \text{ cm}}{5 \text{ cm}} = 0.4$$

For the line that is 10 cm long:

$$\theta = \frac{4 \text{ cm}}{10 \text{ cm}} = 0.4$$

Note that the ratio must have the same units in both the numerator and the denominator. These units will cancel. This ratio without units is known as a **unitless ratio.** It has been given the name **radian,** which means "radius ratio." Therefore,

$$\theta = 0.4 \text{ radian}$$

The direction of rotation, or turn, can be counterclockwise (CCW) or clockwise (CW). **Counterclockwise** means opposite to the direction that the **hands** of a clock turn. **Clockwise** means the direction that the **hands** of a clock turn. If you turn or measure the angle in the CCW direction, it is a positive (+) angle. If you turn or measure the angle in the CW direction, it is a negative (−) angle.

Note that the angle ($\theta = 0.4$ radian) was measured in the counterclockwise direction. Therefore, it is a positive angle.

Study the examples on the following page.

Example

a. Compute the angle ($\Delta\theta$) through which a vector of length **r** = 4 inches turns if the distance that the tip of the vector moves (Δ**s**) is 9.5 inches.

b. Display the vector and the angular distance (angle) turned.

Answer

a. $\Delta\theta = \Delta$**s**/**r**
    = 9.5 inches/4 inches
    = 9.5/4 radians
    = 2.375 radians

b. The vector and the angular distance are shown in the following figure:

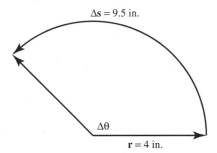

Δ**s** = 9.5 in.

$\Delta\theta$

r = 4 in.

Example

a. Compute the angle (θ) through which a vector of length **r** = 3 cm turns if the distance that the tip of the vector moves (Δ**s**) is 3 cm.

b. Display the vector and the angle turned.

Answer

a. $\Delta\theta = \Delta$**s**/**r**
    = 3 cm/3 cm
    = 1 radian

b. The vector and the angle turned are shown in the following figure:

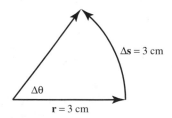

Δ**s** = 3 cm

$\Delta\theta$

r = 3 cm

When the direction of an angle is not specified, the direction of either motion or measure is **CCW.**

Exercise Set 21.1

The answers to the exercises in this chapter are given in Appendix A.

Work the following exercises.

1. **a.** Compute the angle (θ) through which a vector of length $\mathbf{r} = 2$ inches turns if the distance that the tip of the vector moves ($\Delta\mathbf{s}$) is 1.25 inches.

    b. Display the vector and the angle turned.

2. **a.** Compute the angle through which a vector whose length is 5 mm turns if the distance that the tip of the vector moves is 2 cm.

    b. Display the vector and the angle turned. (Be careful of the units!)

3. **a.** Compute the angle through which a vector whose length is 5 inches turns if the distance that the tip of the vector moves is 10 inches CW.

b. Display the vector and the angle turned.

Note: Reasoning will be required for the next two exercises.

4. Compute the distance that the tip of a vector moves if the vector length is 8 inches and the angle turned is 2 radians.

5. Compute the distance that the tip of a vector moves if the vector length is 12 cm and the angle turned is 0.5 radian.

The Measures of Angular Change

There are different measures used to describe angular change. We have introduced one measure of angular change: the **radian.** We shall study this angular measure further, along with three other angular measures: the **revolution,** the **degree,** and the **grad.**

Revolution   The **revolution** is the fundamental measure of angle. The turning vector **r** starts at one point (**A**) and rotates CCW until it returns to that same point. The vector has turned one revolution, as shown in the following figure:

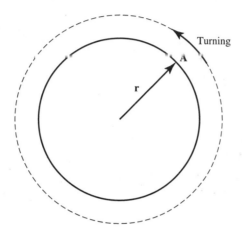

One revolution may be divided into one of the following.

- Two equal parts (semicircles):

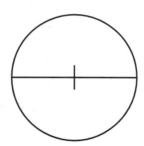

- Four equal parts (**quadrants,** identified by Roman numerals):

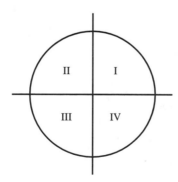

Degree   One revolution may also be divided into 360 equal parts, known as **degrees.** The degree symbol is °. The degree is the most common subdivision of one revolution.

At the end of a revolution, the degree measure may continue or return to zero degrees:

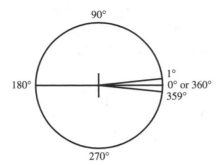

Note the relation between revolutions and degrees:

Revolutions	Degrees (°)
0	0
1/4	90
1/2	180
3/4	270
1	360
2	720

*The **degree** used in angle measure is not related in any way to the **degree** used in temperature measure.*

Radian   We have already introduced the next common measure of one revolution, known as the **radian.** One revolution may be measured as 2π (approximately 6.28) radians.

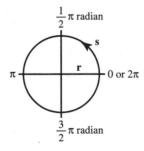

Why is this true? Recall from Chapter 17 that the circumference of a circle with radius **r** is **$2\pi r$.** Now rotate a vector whose length is 1 cm through one revolution. The distance that the tip of this vector moves is

$$\Delta s = 2\pi r = 2\pi \times 1 \text{ cm} = 2\pi \text{ cm}$$

Therefore, the angular change is

$$\theta = \Delta s/r = 2\pi \text{ cm}/1 \text{ cm} = 2\pi \text{ radians}$$

For any circle whose radius is **r,** the angular change is

$$\theta = \Delta s/r = 2\pi r/r = 2\pi \text{ radians}$$

Thus, there are 2π radians in one revolution. This is approximately equivalent to 6.28 radians. Recall from pages 21–4 and 21–8 the sizes of 1 radian and 6 radians. Note that 6 radians is almost equal to 1 revolution.

The degree and the revolution are the more common practical measures of circular distance. However, the **radian** is the most common **math** circular measure.

Note the relation among revolutions, degrees, and radians:

Revolutions	Degrees	Radians
0	0	0
1/4	90	$\pi/2$
1/2	180	π
3/4	270	$3\pi/2$
1	360	2π
2	720	4π

Grad One revolution may also be divided into 400 equal parts, known as **grads.** The **grad** was developed by the French for marking the turning angle of their cannon from the horizontal (zero grads) to the vertical (100 grads). Thus, there are 400 grads in one revolution. In the French military, only the grad measure from zero and to 100 is used. This is considered to be the first quadrant, now calibrated in **grads,** or **graduations.** (We state this fact only because calculators used for computing circular measures have a key marked **DRG,** which represents degrees, radians, and grads.)

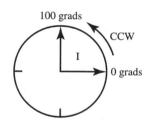

> *In this text, there will be no problems that use the measure **grad.** It is used primarily by the French military. Therefore, if you want to join the French Foreign Legion, then you should first conquer your **math anxieties.** By the time you reach this chapter, you should be doing much better at performing math computations and procedures.*

Recall that we mentioned earlier in this chapter that the measure of angular change, the radian, is a unitless ratio. The same is true of the angle **names** "degree" and "grad"; these names are given to unitless ratios. The names help us separate one method of measuring angular change (or distance) from another method of measuring angular change.

We may now compare these four angular **unitless units** of measure. Here are some of the equivalents:

1 revolution = 360°	1 revolution = 2π radians
360° = 2π radians	¼ revolution = 100 grads
90° = 100 grads	

As noted previously, these angle measures are signed numbers. If you turn or measure an angle in the counterclockwise (CCW) direction, then it is a positive (+) angle. If you turn or measure the angle in the clockwise (CW) direction, then it is a negative (−) angle.

Circular Measure Conversions

We will now use these conversion equivalents and the unity method to convert from one measure of angular change to another.

*It would be helpful if you would now purchase a **protractor** with **degree** markings.*

Study the following examples.

Example

    a. Use the unity method to convert a 90° angle to radian measure.

    b. Construct this angle.

Answer

    a. $90° \times \dfrac{2\pi \text{ radians}}{360°}$

       $= \dfrac{180\,\pi}{360} \text{ radians}$

       $= \dfrac{\pi}{2} \text{ radians}$

    b. The angle is shown in the following figure:

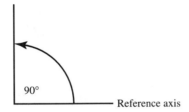

90°   Reference axis

Example

   a. Use the unity method to convert π/3 radians to degree measure.

   b. Construct this angle.

   c. In what quadrant is this angle located?

Answer

   a. $\dfrac{\pi}{3} \text{ radians} \times \dfrac{360°}{2\pi \text{ radians}}$

      $= \dfrac{360}{6} \text{ degrees}$

      $= 60°$

   b. The angle is shown in the following figure:

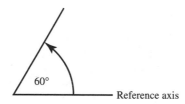

   c. The angle is located in the first quadrant.

Example

   a. Convert one-fifth of a revolution to degrees.

   b. Construct this angle.

   c. In what quadrant is this angle located?

Answer

   a. $\dfrac{1}{5} \text{ revolution} \times \dfrac{360°}{1 \text{ revolution}} = 72°$

   b. The angle is shown in the following figure:

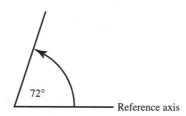

   c. The angle is located in the first quadrant.

Note that 90° is +90°, measured CCW. There are 360° in a circle. Therefore, the equivalent CW angle is

$$90° - 360° = -270°$$

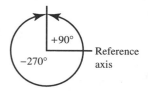

Exercise Set 21.2

Work the following exercises.

1. It is believed that the number **360** was chosen for the number of degrees in a circle because it contains so many prime factors. Compute the prime factors of 360.

2. **a.** Convert a 45° angle to radian measure.

    b. Construct this angle.

    c. In what quadrant is this angle located?

3. **a.** Convert a $\pi/6$ radian angle to degree measure.

    b. Construct this angle.

    c. In what quadrant is this angle located?

4. **a.** Convert one-sixth of a revolution to degrees.

    b. Construct this angle.

    c. In what quadrant is this angle located?

5. **a.** Convert two-thirds of a revolution to degrees.

   b. Construct this angle.

   c. In what quadrant is this angle located?

6. Convert two-thirds of a revolution to radians.

7. The 180° angle is a special angle.

   a. Convert a 180° CCW angle to a CW angle.

   b. Construct these angles.

8. Convert a 180° angle to radian measure.

Note that turning a vector by either +180° or −180° is equivalent to reversing the direction of that vector. This fact is often used in science and engineering computations.

Reducing Angles Greater than One Revolution

A line, vector, or other object may turn through more than one revolution. Therefore, angles can exceed 360°, one revolution or 2π radians, in both the positive and the negative direction.

Scientific calculators are internally programmed to reduce angles greater than 360° or 2π radians to an equivalent positive angle that is less than 360° or 2π radians. If you do not have a simple scientific calculator, you should purchase one now. You must have it to solve some of the problems that involve trigonometry, which we will discuss later in this chapter.

> *The minimum keys that you will need on a scientific calculator are listed in Appendix G. Scientific calculators may also have a **DRG** key. As mentioned earlier, this key is used to convert the operation of the calculator from degrees (D) to radians (R) to grads (G).*

Study the following example.

Example   A point on a wheel turns through a total of 1140°.

    a. Where is this point located from the reference axis?

    b. Construct this angle.

    c. In what quadrant is this angle located?

Answer

    a. Continue to subtract 360° until you reach a number that is less than 360°:

$$1140° - 360° = 780°$$
$$780° - 360° = 420°$$
$$420° - 360° = 60°$$

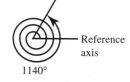

    b. The angle is shown in the following figure:

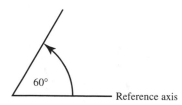

    c. The angle is located in the first quadrant.

Exercise Set 21.3

Work the following exercises.

1. A line turns through a total of 820°.

    a. Where is this line now located from the reference axis?

    b. Construct this angle.

    c. In what quadrant is this angle located?

   *Be careful when working the next problem. You will be working with a clockwise angle. Construct your large negative angle first, and then **reason** where the line will stop. How does this affect your math calculations?*

2. A line turns through a total of −680°.

    a. Where is the line now located from the reference axis?

    b. Construct this angle.

    c. In what quadrant is this angle located?

3. A line turns through a total of 1450°.

   a. Where is this line now located from the reference axis?

   b. Construct this angle.

   c. In what quadrant is this angle located?

4. A line turns through a total of −1800°.

   a. Where is this line now located from the reference axis?

   b. Construct this angle.

   c. In what quadrant is this angle located?

An Introduction to Trigonometry

It is not always possible to directly measure a distance. The use of angles and special formulas can assist us in accurately determining distances that are difficult, or otherwise impossible, to measure.

Study the following example.

Example   The drawings that contain the exact height of a tall building have been lost. An engineer needs to accurately determine the height of this building. As shown in the following figure, a surveyor stands exactly 250 ft from the corner of the building. He measures the angle from the horizontal to the top of the building. The angle is 53.24°. What is the height of this building?

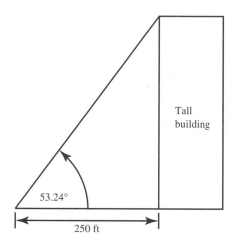

The solution to this example is given on page 21–21. First, we will discuss the methods and calculations involved in solving this type of problem.

In approximately A.D. 1460, Johann Muller described the solution to problems such as this in a geometry book. In approximately A.D. 1500, Nicolaus Copernicus applied many of these concepts and formulas to studying the paths of the planets around our sun.

All geometry formulas are related to the formula discovered by Pythagoras. Therefore, we shall start with a review of the **right triangle.** We now know that the right angle of a right triangle is an angle of exactly 90° or $\pi/2$ radians:

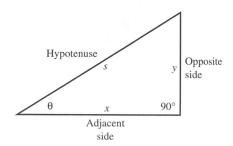

If the side adjacent to angle θ is also horizontal, then it can be shown on the Cartesian coordinate *x*-axis.

If the side opposite angle θ is vertical, then it can be shown on the *y*-axis.

Therefore, the angle between the opposite and adjacent sides must be a right angle (90°). The side opposite the right angle is known as the **hypotenuse.**

If you examine all the combinations of ratios of these three sides, then you will discover that there are six possible combinations. Only three of these ratios are of interest to us at this time. (Note that the word **side** is omitted.)

- The **sine** of θ:

$$\sin \theta = \frac{\text{opposite}}{\text{hypotenuse}} = \frac{y}{s}$$

- The **cosine** of θ:

$$\cos \theta = \frac{\text{adjacent}}{\text{hypotenuse}} = \frac{x}{s}$$

- The **tangent** of θ:

$$\tan \theta = \frac{\text{opposite}}{\text{adjacent}} = \frac{y}{x}$$

Even though the abbreviations are three-letter words, they are spoken "**sine,**" "**cosine,**" and "**tangent.**"

These three ratios must always involve the same units of measure. Therefore, these ratios are unitless ratios.

Because these are ratios, their numerical value depends upon the length of two sides. Therefore, if the two lengths change by the **same proportion,** these ratios will remain constant. Thus, the lengths will have the same ratio value.

These three fundamental definitions are the fundamentals of trigonometry. **Trigonometry** is the name given to the branch of mathematics that involves the study of triangles, their sides, and their angles.

There are three more ratios used in trigonometry:

- The reciprocal of the sine is known as the **cosecant (CSC).**
- The reciprocal of the cosine is known as the **secant (SEC).**
- The reciprocal of the tangent is known as the **cotangent (COT).**

The names of these reciprocal definitions do **not** appear on hand-held calculators. We shall not use them in this text. In general, they are of interest to persons who wish to become more involved in trigonometry.

The sine, cosine, and tangent ratios lead directly to the following six formulas. Again, the word **side** is omitted:

$$\text{Opposite} = (\text{hypotenuse}) \times (\sin \theta)$$

$$\text{Opposite} = (\text{adjacent}) \times (\tan \theta)$$

$$\text{Adjacent} = (\text{hypotenuse}) \times (\cos \theta)$$

$$\text{Adjacent} = \frac{\text{opposite}}{\tan \theta}$$

$$\text{Hypotenuse} = \frac{\text{opposite}}{\sin \theta}$$

$$\text{Hypotenuse} = \frac{\text{adjacent}}{\cos \theta}$$

In these formulas, two of the three **quantities** must be known. The third quantity will be the unknown to be calculated; it should be on the left-hand side of the **equal (=)** sign.

How do you select which one of these six formulas you should use to solve your problem? First, determine the quantity you want to calculate. Second, select the formula that has the quantity you want to calculate on the left-hand side of the **equal (=)** sign and your two known quantities on the right-hand side of the **equal (=)** sign.

On the following page, we continue with the previous example.

Example The drawings that contain the exact height of a tall building have been lost. An engineer wants to accurately determine the height of that tall building. A surveyor stands exactly 250 ft from the corner of the building. He measures the angle from the horizontal to the top of the building. If the angle measures 53.24°, what is the height of this building?

Answer Construct the right triangle. Mark the knowns and unknowns on this triangle, as shown in the following figure:

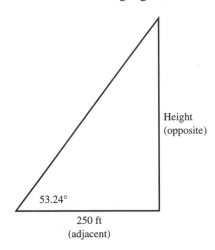

The two knowns are the angle (53.24°) and the side adjacent to that angle (250 ft). The unknown is the height, which is the side opposite the known angle.

Select the formula

$$\text{Opposite} = (\text{adjacent}) \times (\tan \theta)$$

This formula now becomes

$$\begin{aligned} \text{Height} &= (250 \text{ ft}) \times (\tan 53.24°) \\ &= (250 \text{ ft}) \times (1.338675) \\ &= 334.67 \text{ ft} \end{aligned}$$

Before you begin to enter angle values, be certain that your calculator is in the degree (**D** of the **DRG**) mode.

Your scientific calculator may require that you enter 53.24 before you depress the TAN key. Test this by entering 53.24 and then depressing the TAN key. If the result is a value close to 1.338675, then you should calculate using your trigonometry formulas from right to left as follows:

Step 1: Enter **53.24.**
Step 2: Depress **TAN.**
Step 3: Depress **×.**
Step 4: Enter **250.**
Step 5: Depress **=.**

Your result should be a value close to 334.67.

Study the example on the following page.

Example   A surveyor has a job measuring the distance across a river, that is, its **width.** On the opposite side of the river are two poles that are 300 ft apart. The surveyor places her equipment so that her **line of sight** to one pole forms a right angle to an imaginary line between the two poles. The angle to the other pole is measured to be 27.83°.

a. Construct a sketch of the problem; then construct the applicable triangle.

b. Determine the width of the river.

Answer

a. The problem is sketched and the triangle is drawn in the following figures:

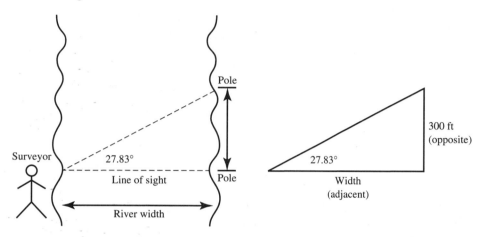

b. The two knowns are the angle (27.83°) and the side opposite that angle (300 ft). The unknown is the river width, which is the side adjacent to the known angle. Select the formula

$$\text{Adjacent} = \frac{\text{opposite}}{\tan \theta}$$

This formula now becomes

$$\text{Width} = \frac{300 \text{ ft}}{\tan 27.83°} = \frac{300 \text{ ft}}{0.5279095}$$

$$= 568.3 \text{ ft across the river}$$

This example involves division by a trigonometric ratio. Perform the computation on your calculator as follows:

Step 1:	Enter **300.**
Step 2:	Depress ÷.
Step 3:	Enter 27.83.
Step 4:	Depress **TAN.**
Step 5:	Depress **=.**

*Your result should be a value close to **568.3.***

CIRCULAR MEASURE

Exercise Set 21.4

Work the following exercises; also prepare sketches.

1. A surveyor wants to know the height of a building. The distance from the bottom of that building to a surveyor on the ground is 410 ft. The angle between the horizontal and the top of the building is 59.3°. How high is the building?

2. The height of a building is known to be 330 meters. The distance from the bottom of that building to a surveyor at ground level is 78 meters. What is the angle between the surveyor at ground level and the top of the building?

3. The distance between two right-angle bridge strut fasteners, shown in the following figure as circles, is to be 50 meters at an angle of 42°:

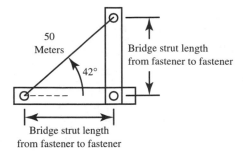

What must be the length of each of the two bridge struts from fastener to fastener?

The computed distances (the opposite and adjacent sides) are from fastener to fastener. The fasteners might be bolts, shown as circles. The actual strut lengths require adding additional amounts of length beyond the bolts.

4. For radio and television signals, the **line-of-sight distance** is often very important. As shown in the following figure, the top of one radio tower is 117 ft above sea level (a reference). The top of a second radio tower, located on a mountain, is 2423 ft above sea level. The angle from one tower to the second tower is measured to be 36.52° from the horizontal:

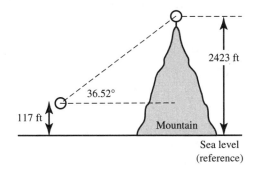

What is the line-of-sight distance between the top of the two radio towers?

Summary

Angles are used to measure **turning.** A line, vector, or point on a wheel turning in a circular pattern generates an angle.

The **reference axis** for most angular measure is the positive *x*-axis. The angular change $\Delta\theta$ or θ is defined as the ratio of two linear distances (Δs and **r**):

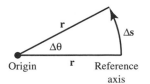

$$\Delta\theta = \frac{\Delta s}{r}$$

Thus, an **angle** is the ratio of two distances; it is a unitless ratio. However, the **value** of $\Delta\theta$ is not dependent upon the lengths of **r** and Δs; it is dependent upon their **ratios.**

The positive (+) direction of angular measure is **counterclockwise (CCW)** from the reference position. The negative (−) direction of angular measure is **clockwise (CW)** from the reference position.

Angles are given using one of four measures of angular change or distance. In order to distinguish among each of these measures, it has become necessary to **name** these unitless ratios. The four names are the **revolution, degree, radian,** and **grad.**

One **revolution** is the distance a line or vector travels as it turns from a reference point and an axis back to the position from which it started:

> There are 360° in one revolution.
>
> There are 2π **radians** in one revolution.
>
> There are 100 **grads** in one-quarter of a revolution.

The grad is used mostly by the French military.

Vectors, lines, and wheels can turn through more than one revolution. For angular measure, each revolution is identical to every other revolution. An angle (such as 400°) may be written more simply as an angle that is less than 360°. For example:

$$400° - 360° = 40°$$

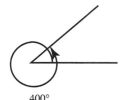

400°   40°

Trigonometry is the name of the branch of mathematics that includes a description of triangles, their sides, and their angles. Three fundamental right-triangle ratios have been defined as follows:

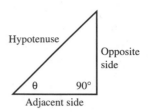

$$\sin \theta = \frac{\text{opposite}}{\text{hypotenuse}}$$

$$\cos \theta = \frac{\text{adjacent}}{\text{hypotenuse}}$$

$$\tan \theta = \frac{\text{opposite}}{\text{adjacent}}$$

These ratios are unitless ratios.

Six formulas can then be derived from these right triangle definitions:

$$\text{Opposite} = (\text{hypotenuse}) \times (\sin \theta)$$

$$\text{Opposite} = (\text{adjacent}) \times (\tan \theta)$$

$$\text{Adjacent} = (\text{hypotenuse}) \times (\cos \theta)$$

$$\text{Adjacent} = \frac{\text{opposite}}{\tan \theta}$$

$$\text{Hypotenuse} = \frac{\text{opposite}}{\sin \theta}$$

$$\text{Hypotenuse} = \frac{\text{adjacent}}{\cos \theta}$$

Thus, if one side and either angle are known, then the other two sides may be computed.

The cosecant, secant, and cotangent are also used in texts that focus on trigonometry. They do **not** appear on a calculator. Therefore, we have not included a discussion of them in this chapter.

Glossary of Math Words

Angle (an′-gl)   A measure of the amount that a line or a vector turns around a fixed point; the fixed point is the center.

Arc (ark′)   A portion of the circumference of a circle.

Clockwise (klok′-wyze) **(CW)**   A turning in the same direction as the **hands** of a clock.

Cosine (ko′-syne) **of an angle**   The name given to one ratio of two sides of a right triangle; it is the side adjacent to the angle divided by the hypotenuse of a right triangle.

Counterclockwise (koun′-tur-klok′-wyze) **(CCW)**   Turning in the opposite direction to the **hands** of a clock.

Degree (di-gree′)   An angular measure, symbolized by °; there are 360° in one revolution.

Delta theta (del′-ta thay′-ta)   A change in the angle of a turning line or vector, measured from its starting (first) position to its ending (second) position; symbolized $\Delta\theta$.

Grad (grad′)   An angular measure; there are 100 grads in one-fourth of a revolution, or one quadrant.

Line of sight (L.O.S.) (lyne′ of site′)   An imaginary straight line joining the center of the eye of the observer with the object viewed.

Quadrant (kwad′-rant)   One-fourth (quarter) of a circle or revolution; the four quadrants are identified using the Roman numerals I, II, III, and IV.

Radian (ray′-dee-un)   An angular measure; there are 2π radians in one revolution.

Reference axis (ref′-ur-ense ak′-sis)   The axis, or line, that is the starting place for a turning line or vector; it is often the positive horizontal (x) axis.

Revolution (rev′-uh-loo′-shun)   The fundamental angular measure; it is the angular distance measured from a starting line that rotates around the center of a circle until it returns to that same starting line.

Sine (syne′) **of an angle**   The name given to one ratio of the sides of a right triangle; it is the side opposite the angle divided by the hypotenuse of a right triangle.

Tangent (tan′-jent) **of an angle**   The name given to one ratio of the sides of a right triangle; it is the side opposite the angle divided by the side adjacent to the angle of a right triangle.

Theta (thay′-ta)   The Greek letter θ commonly used to identify an angle.

Trigonometry (trig-uh-nom′-uh-tree)   The branch of mathematics that includes a description of triangles, their sides, and their angles.

Chapter 21 Test

Follow instructions carefully:

> *On a separate piece of paper, write the answers to the following questions. Do not write on these pages.*
>
> *When you are finished, compare your answers with those given in Appendix B.*
>
> *Record the date, your test time, and your score on the chart at the end of this test.*

The formula that describes the distance (Δs) that the tip of a vector of length **r** turns as it moves through an angle ($\Delta\theta$) is

$$\Delta s = r\,\Delta\theta \quad \text{where } \Delta\theta \text{ must be in radian measure}$$

1. Compute the distance that the tip of a vector moves if the vector length is 9 inches and the angle turned is $\pi/3$ radians.

2. Compute the distance that the tip of a vector moves if the vector length is 21 mm and the angle turned is $90°$.

Angular change ($\Delta\theta$) is defined as the ratio of the distance (Δs) that the tip of a vector turns to the length (**r**) of the vector:

$$\Delta\theta = \Delta s/r$$

3. Compute the angular change in radians if the distance that the tip of a vector moves is 60 cm and the length of the vector is 20 cm.

4. Convert the angular change in Problem 3 to degrees.

5. Convert a 210° angle to radian measure.

6. In what quadrant is the vector of Problem 5 located?

7. Convert a $2\pi/3$ radian angle to degree measure.

8. In what quadrant is the vector of Problem 7 located?

9. Convert three-quarters of a revolution to degrees.

10. Convert three-quarters of a revolution to radians.

11. A wheel turns 3.5 revolutions. Convert this angle to a value less than 360°.

12. A wheel turns 3.5 revolutions. Convert this angle to a value less than 2π radians.

13. Convert a 660° angle to radian measure.

14. In what quadrant is the vector of Problem 13 located?

15. Convert a $7\pi/2$ radian angle to revolution measure.

16. A vector turns through a total of $-1105°$. Where is this vector now located from the reference axis?

Use the following right triangle for the computations required in the next two problems:

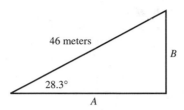

17. For the given triangle, compute the value of the side marked *A*.

18. For the given triangle, compute the value of the side marked *B*.

Use the following right triangle for the computations required in the next two problems:

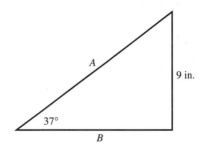

19. For the given triangle, compute the value of the side marked *A*.

20. For the given triangle, compute the value of the side marked *B*.

Chapter 21 Test Record

DATE	TIME	SCORE

Refer to

- Appendix B for the correct answers to this test.
- Appendix C if taking this test required too much effort.
- Appendix D for **Memory Methods** assistance.
- Appendix E if your test scores are decreasing.

When a problem seems difficult, find one like it in this chapter. Then study that (and the related material) again.

Develop additional 3 × 5 cards for those ideas, problems, and procedures that caused you difficulty.

Note: The Part 7 Review Test begins on page 21–34.

Part 7 Review Test

CHAPTER 20
Sequences and Series

CHAPTER 21
Circular Measure

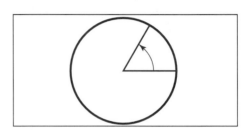

Instructions

Review your 3 × 5 cards before you take this test.

Complete this test.

You may find some ideas, problems, or procedures difficult or confusing. Develop 3 × 5 cards for them.

Part 7 Review Test

1. For the sequence 5, 13, 21, 29, determine the next term.

2. For the sequence 5, 13, 22, 32, determine the next term.

3. Determine the common difference for the progression 32, 38, 44, 50.

4. For the sequence 6, −9, 12, −15, determine the value of the series.

5. Determine the common ratio for the progression 3, 12, 48, 192.

6. A sequence consists of six terms. The first four terms are 1, 3, 6, 10. Determine the fifth and sixth terms.

The first term of a five-term geometric progression is 4. The second term of this progression is 12.

7. What is the common ratio?

8. Determine the value of the third, fourth, and fifth terms.

Many economists continually inform us that "a little inflation won't hurt you." Therefore, you decide to save money for your retirement. Assume that the "cost of living" increases at the rate of 5% per year.

9. How much will the accumulated inflation become at the end of twenty years?

10. How long will it take the value of your saved dollar to halve? (This is another way of stating, "How long will it take the accumulated inflation to reach 200%?")

11. Compute the distance that the tip of a vector moves if the vector length is 40 cm and the angle turned is $\pi/4$ radians. (This distance is a portion of the circumference of a circle.)

12. Convert a 150° angle to radian measure.

13. Convert a $3\pi/2$ radian angle to degree measure.

14. Convert two-fifths of a revolution to degrees.

A wheel turns 2.5 revolutions. Convert this angle into the following degrees.

15. greater than 360°

16. between 0° and 360°

17. Convert a $7\pi/6$ radian angle to degree measure.

18. Convert five-eighths of a revolution to radian measure.

A line turns through a total of $13\pi/6$ radians. Convert this angle into the following degrees.

19. greater than 360°

20. between 0° and 360°

Part 7 Review Test Record

DATE	TIME	SCORE

Refer to

- Appendix B for the correct answers to this test.
- Appendix C if taking this test required too much effort.
- Appendix D for **Memory Methods** assistance.
- Appendix E if your test scores are decreasing.

When a problem seems difficult, find one like it near the page referenced with the answers in Appendix B. Study the related material.

Develop additional 3×5 cards for those ideas, problems, and procedures that caused you difficulty.

Word Problems

Some Words about Word Problems

In our everyday world, most problems that we are expected to solve are expressed in words. Someone may ask us a question that can be solved using math. For example, a supervisor may tell us to analyze some data. We must find and apply the appropriate math, but first we must translate English words into math numbers and operations.

The English language is not a precise language. Attempts by scholars and by teachers of English to make this language more precise have seldom worked. Writers, reporters, television announcers, script writers, and we users of English have often been very sloppy in its use. As a result, translation from English words into math statements is difficult at times.

You have completed many **word problems** in **every** previous chapter. We shall next review those types of problems and, where possible, group them into types or **families** of word problems. We have chosen the wording of these problems to be as precise as the English language and our writing skills have permitted.

Six Formal Steps to Solving Word Problems

The previous chapters have been preparing you so that you can now solve word problems more easily. Let us become more formal and explore six steps often used in the solution of word problems.

Step 1:	Read the words.
Step 2:	Determine the math required.
Step 3:	Construct and label a sketch where necessary.
Step 4:	Translate the words into math.
Step 5:	Calculate your result(s).
Step 6:	Compare your result(s) with the wording of the problem.

These steps have been carefully chosen. They can also be used if you decide to study word problems that require the use of algebra, geometry, trigonometry, calculus, or statistics.

We will examine these six steps, one step at a time.

Step 1 *Read the words.*

Read and reread the problem until you understand its meaning. You may have to read the problem several times. You must understand the problem well enough so that you can:

a. Identify the **given** or **known** values.

b. Identify the **unknown** value(s) or **solution** requested.

c. Write two separate lists: knowns and unknowns.

Step 2 *Determine the math required.*

Determine the fundamental math operations required to solve the problem. You may find the following chart helpful:

The Math Operation	Is Indicated by:	
	The Notation	**The Word or Phrase**
Addition	2 +	and more than increased by sum
Subtraction	2 −	less than decreased by difference
Multiplication	2 × 2 × 3 × n ×	product twice doubled tripled n times
Division	÷ 2 ÷ 3 ÷ 4	quotient halved; one-half of one-third of quartered; one-fourth of

As you study the examples in this chapter and the next, you may discover other words that **match** these four fundamental math operations. Write them in this chart in the appropriate place. Use a pencil so that you can erase them if you later find that they do not always **fit** your needs.

For some word problems, the math required is easily identified. For more complicated word problems, some of the problem may have to be solved before all of the math required can be determined.

Step 3 *Construct and label a sketch where necessary.*

If geometry or trigonometry is involved, then construct and label a sketch. A **picture** is often worth more than one thousand words when word problems are involved. It is an important intermediate step in the translation of words into math, as the following example illustrates.

Example What are the area and the perimeter of a rectangle that has a length of 5 cm and a width of 3 cm?

Answer Read the words of the problem carefully, and consider the known values (rectangle length = 5 cm and rectangle width = 3 cm) as well as the unknown values (area and perimeter). Then construct a sketch of this rectangle:

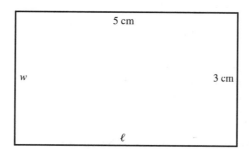

Examine your list of formulas and select the two formulas that apply to this problem:

$$\text{Area} = \ell w$$
$$\text{Perimeter} = 2\ell + 2w$$

We will return to the answer in the next step.

The unknown must always be on the left-hand side of the formula; the knowns must always be on the right-hand side of the formula.

Why is the unknown always on the left-hand side in our formulas? Because all of the formulas are presented in a simple and consistent manner! While there is no math requirement that limits the location of the unknown and knowns, we recommend this method of presentation.

Step 4 *Translate the words into math.*

Return to the word problem of Step 3. Translate the English words into math expressions using the appropriate formulas and the fundamental math operations.

Answer, Cont.

$$\text{Area} = \ell w$$
$$\text{Perimeter} = 2\ell + 2w$$

In this example, the length of the rectangle is 5 cm, and the width of the rectangle is 3 cm. Therefore,

$$\text{Area} = (5 \text{ cm}) \times (3 \text{ cm})$$
$$\text{Perimeter} = 2 \times 5 \text{ cm} + 2 \times 3 \text{ cm}$$

Step 5 *Calculate the results.*

Perform the math operations indicated.

Include any units of measure where they apply.

Answer, Cont.

$$\text{Area} = 15 \text{ cm}^2 \quad \text{or} \quad 15 \text{ sq cm}$$
$$\text{Perimeter} = (2 \times 5) \text{ cm} + (2 \times 3) \text{ cm}$$
$$= 10 \text{ cm} + 6 \text{ cm}$$
$$= 16 \text{ cm}$$

Step 6 *Compare the result(s) with the wording of the problem.*

Return to the wording of the problem. The solution and the wording should agree.

Answer, Cont.   Did you solve for the unknowns: the area and the perimeter of the rectangle? Do the final units of sq cm and cm **fit** the units of area and perimeter? Yes, they do in this solution.

As you gain experience, you will also be able to see that the values of **15** and **16** are reasonable results.

Two Applications of the Six Steps

Study the following examples.

Example   A farmer has prepared some land for a rectangular garden. The land measures 40 ft by 50 ft. The garden has a fence around it and requires 60 pounds of fertilizer each spring.

The farmer decides to remove trees and brush and to increase the garden until it measures 80 ft by 100 ft. How much more fence must be purchased, and how much fertilizer will be needed next spring?

Answer

1. *Read the words.*

   There seem to be two problems combined into one:

   Old garden: 40 ft × 50 ft with 60 lb of fertilizer

   New garden: 80 ft × 100 ft with **?** lb of fertilizer

Knowns

Old garden dimensions

New garden dimensions

Old garden fertilizer needed

Unknowns

Old garden perimeter and area

New garden perimeter and area

New garden extra fence required

New garden total fertilizer needed

2. *Determine the math required.*

There appear to be at least addition (perimeter) and multiplication (area) required. After you have solved this part of the problem, there may be more math required.

3. *Construct and label a sketch where necessary:*

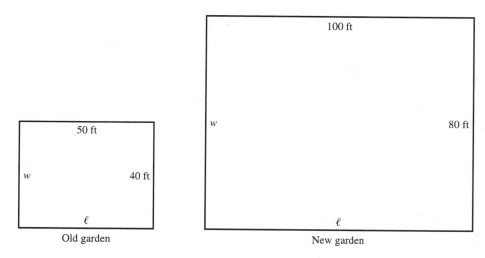

Old garden   New garden

For both gardens,

$$\text{Perimeter} = 2\ell + 2w$$

$$\text{Area} = \ell w$$

Now determine the two garden perimeters and areas. Wait until later to determine the other results and answer the problem questions.

4. *Translate the words into math.*

Old Garden

Perimeter = 2 × (50 ft) + 2 × (40 ft)
        = 100 ft + 80 ft = 180 ft

Area = (50 ft) × (40 ft) = 2000 ft²

New Garden

Perimeter = 2 × (100 ft) + 2 × (80 ft)
        = 200 ft + 160 ft = 360 ft

Area = (100 ft) × (80 ft) = 8000 ft²

5. *Calculate the results.*

Some of the **results** were calculated in Step 4. However, the two questions have not been answered. Use **reasoning** to determine the amount of extra fence needed.

> Extra fence needed
> = new perimeter − old perimeter
> = 360 ft − 180 ft = 180 ft

The **fertilizer** computation may not be as simple. Compare the new garden area to the old garden area. It is

$$\frac{\text{New garden area}}{\text{Old garden area}} = \frac{8000 \text{ ft}^2}{2000 \text{ ft}^2} = \frac{4}{1}$$

Note that the new garden will require **four times** as much fertilizer as the old garden required.

> New fertilizer needed
> = 4 × 60 lb = 240 lb

6. *Compare the results with the wording of the problem.*

How much more fence will be needed? Note that the extra 180 feet of fence is the same amount as the fence around the old garden. The fence requirement has doubled. Examine the two pictures; it is true. How much fertilizer will be needed next spring? Examine the two pictures again. The new garden may be divided into four equal **old** gardens.

Example   There are two households on a rural road. The first household consists of five people whose ages are 18, 19, 44, 69, and 65. The second household consists of four people whose average age is 28. What is the average age of the people in the first household? What is the average age of both households together?

Answer

1. *Read the words.*

There seem to be two problems; list the knowns and the unknowns.

Knowns

Number of people in the first household

Ages of people in the first household

Number of people in the second household

Average age of people in the second household

Unknowns

Average age of people in first household

Ages of people in second household

Average age of people in both households

2. *Determine the math required.*

There appear to be at least addition and division required because this is a problem that requires averaging. After you have solved part of the problem, there may be more math required.

3. *Construct and label a sketch where necessary.*

A sketch is not necessary.

4. *Translate the words into math.*

Examine the first question: What is the average age of the people in the first household? Translate these words into math.

> Average age for first household
> = sum of all ages in household 1 divided by
>   the number of persons in household 1
> = (18 + 19 + 44 + 69 + 65) ÷ 5
> = 215 ÷ 5
> = 43

5. *Calculate the results.*

Some of the **results** were calculated in Step 4. However, the second question has not been answered.

Examine the second question: What is the average age of both households together? How can we determine this result when we don't know what the individual ages are for the people in the second household?

Recall how we computed the average age for the first household. First, we determined the sum of all ages, and then we divided that sum by the number of people in that household.

Do we know the sum of all the ages in household 1 and household 2? Yes, the sum of the five ages in household 1 is 215, and the sum of the four ages in household 2 is 28 × 4. This becomes

> 215 + 28 × 4
> = 215 + 112
> = 327

Now compute the average age of both households together. Divide this sum (327) by the total number of people (5 + 4):

> Average age for both households
> = sum of all ages in both households divided by
>   the number of persons in both households
> = (327) ÷ (4 + 5)
> = 327 ÷ 9
> = 36.3 . . . years of average age for both households

6. *Compare the results with the wording of the problem.*

The average age for the first household is 43.

The average age for the second household is 28 (given).

The average age for both households together is 36.3. . ..

Note that the average age for both households is between the values for each household separately, which makes sense because they have an almost equal number of people.

Practice Problems

The remainder of this chapter consists of practice word problems for you to solve in the Exercise Sets. The first several problems are simple in wording. They will provide you the opportunity to review and strengthen the knowledge you gained from the previous chapters. More complicated word problems will be introduced gradually.

No person has been able to solve every word problem written. Some math scholars have devoted their lives trying to solve certain types of word problems. This book avoids these problems.

You should now have enough knowledge and reasoning ability to solve all of the problems in this chapter. Where you might have some difficulty, comments *in italic type* are provided.

As the word problems become more difficult, carefully and methodically apply the six steps previously described and demonstrated:

1. Read the words.
2. Determine the math required.
3. Construct and label a sketch where necessary.
4. Translate the words into math.
5. Calculate the result(s).
6. Compare the result(s) with the wording of the problem.

Your life experiences may not have included some of the topics of the more complicated word problems. If you need to, ask a friend for help.

Exercise Set 22.1

The answers to the exercises in this chapter are given in Appendix A.

Work the following exercises. They are similar; the first one requires the direct application of the **mean average** definition.

1. There are two offices in one building. The ages of the six persons in the smaller office are

$$20, \quad 31, \quad 22, \quad 29, \quad 24, \quad 30$$

The ages of the twelve persons in the larger office are

$$19, \quad 42, \quad 24, \quad 29, \quad 36, \quad 27, \quad 25, \quad 53, \quad 34, \quad 46, \quad 38, \quad 47$$

   a. What is the average age of the persons in the smaller office?

   b. What is the average age of the persons in the larger office?

   c. What is the average age of the persons in the building?

Exercise Set 22.1, *Continued*

This second exercise also requires that you understand the **mean average** definition and that you be able to apply that definition to a slightly different type of problem.

2. There are two classrooms in a small private school. In the classroom that has fifteen lower-grade students, the average age is 8.2 years. In the classroom that has twenty-one upper-grade students, the average age is 13 years. What is the average age of the students at this private school?

    At first you may believe that you do not have enough information to solve this exercise. You have exactly enough information. Compare this exercise with Exercise 1. Study how you solved that problem. Then you should be able to solve this problem.

Exercise Set 22.1, *Continued*

This third exercise is a more complicated version of the previous one.

3. The metropolitan area of Blatt consists of an inner city and six suburbs. Government census data are available for each of the seven groupings as follows:

Census for	Population	Average Age
Blatt City	1.6 million	37.6
North Blatt	250 thousand	46.1
South Blatt	352 thousand	32.4
East Blatt	145 thousand	57.3
West Blatt	421 thousand	45.2
Durango	85 thousand	39.6
Weditgass	129 thousand	51.7

   a. What is the average age of the population of the six suburbs?

   b. What is the average age of the entire population of the metropolitan area of Blatt?

Exercise Set 22.2

The next three exercises require the computation of surface area. Units of measure will change within an exercise. Carefully write your explanation of how you solved each of these exercises. This explanation is known as the **documentation** of your solution and is increasingly a very important requirement of most industrial companies throughout the world.

1. A rectangular table is 13 ft long and 7 ft wide. It is to be covered with cloth that is to hang over each edge of the table by 1 foot.

    a. How many square feet is the tablecloth?

    b. Material is available in the following widths: 35″, 44″, 58″, and 60″. Ignoring seams and hems, how many square feet of cloth must be purchased? (*Hint:* This size may be different from the size of the tablecloth.)

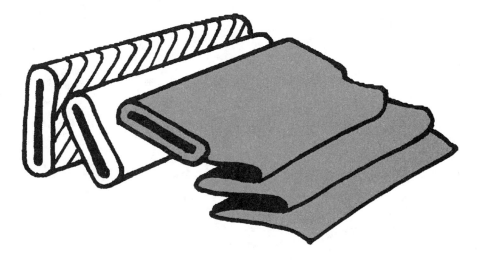

    c. If the cloth costs $3.50 per square yard, then what will be the cost of the cloth?

    *In business, the **$3.50 per square yard** is often (incorrectly) written **$3.50 per yard.** It is also sold by the **running yard,** which means **one yard long** and whatever widths are available, such as 35″, 44″, 58″, or 60″.*

Exercise 22.2, *Continued*

2. A display consists of a 2-foot-square board. Two 14-inch-diameter circular pieces are to be mounted *flat* on the board, one on each surface. These three pieces of the display will be constructed of new, unpainted wood. Before the pieces are assembled, both surfaces of the board, and one surface of each circle, are to be painted with a special enamel paint. The circular pieces will then be mounted on the board, with the unpainted side of each circle exposed. Then a second coat of paint is to be applied to the visible portions of both surfaces of the square board only.

   The special enamel paint is sold by the pint can. The instructions on the can state: **This can will cover 8 square feet for the first coat, and 12 square feet for each additional coat. Allow 4 hours of drying time between applications.**

   a. How many cans of paint will be needed?

   b. The paint costs $8.50 per pint if only one can is purchased. A **sale** sign states that the paint will cost $15.50 per quart if two or more cans are purchased at the same time. How much will the paint cost if all of the cans are purchased at the same time?

   Conversion

   2 pt = 1 qt

Exercise Set 22.2, *Continued*

3. Two identical table tops are constructed; each is circular in shape and has a radius of one yard. The center portion of the first table top has a 14-inch-diameter round hole cut in it.

   a. How much table top remains to be painted?

   b. The second table top has a square-hole, 15 centimeters on a side, cut in it. How much table top remains to be painted?

   Conversions
   3 ft = 1 yd
   12 in. = 1 ft
   2.54 cm = 1 in.

Exercise Set 22.3

The next three exercises examine distance traveled and the rate (speed or velocity) at which that distance is traveled. Units of measure will be involved.

1. The dimensions of a rectangular barn are 40 ft by 80 ft. A goat is tied to one corner of the outside of the barn using a 10-foot rope. The goat is standing next to one side of the barn, with the rope stretched as much as possible (10 ft). The goat then walks, keeping the rope stretched all of the time, to the other side of the barn.

   a. How far has the goat walked?

   b. The goat then notices that all of the ground it can reach is covered with tasty grass. How many square feet of tasty grass is available for the goat to eat?

Exercise Set 22.3, Continued

2. A runner from the United States is to compete with a runner from Romania. The runner from the United States requires 0.92 minute to run 1200 feet. The runner from Romania requires 62 seconds to run 400 meters.

    a. What is the speed of each runner in miles per hour?

    b. What is the speed of each runner in meters per second?

    Conversions
    1 mi = 5280 ft
    1 ft = 12 in.
    1 in. = 2.54 cm
    1 m = 100 cm

Exercise Set 22.3, *Continued*

3. A trip is planned; driving will occur in both the United States and Canada. Maps of both nations are studied. The reference point will be at Oshkosh in the United States. The distances to be traveled from Oshkosh to Ipsum (in Canada) are

150 mi north,   60 mi east,   25 mi north,   40 mi south,
55 km north,   125 km west,   40 km north,   250 km west,
95 km south,   50 km west,   10 km south

   a. What is the total distance traveled in miles?

   b. What is the total distance traveled in kilometers?

The following records of gas purchased between Oshkosh and Ipsum were recorded:

7.8 gal,   6.1 gal,   36.2 L,   29.7 L

   c. What is the total gas consumption in miles per gallon?

   d. What is the total gas consumption in kilometers per liter?

Conversions

$1 L = 0.264$ gal

1 mi $= 5280$ ft

1 ft $= 12$ in.

1 in. $= 2.54$ cm

1 m $= 100$ cm

1 km $= 1000$ m

Exercise Set 22.4

The next three exercises are **number** problems. You may discover that these problems require more **thinking** than writing. Number problems are abstract; diagrams are usually limited to a display on the real number line or to a chart of values.

1. The odd integers start at **1**.

    a. List the first seven (odd) integers.

    b. Display these integers on the real number line.

    c. Compute the series for this seven-number sequence.

    d. What is the eighty-second odd integer?

    e. What is the sum of all the odd integers from **1** through (including) the eighty-seventh?

Exercise Set 22.4, *Continued*

2. Dan is 46 years old. He says, "I am ten years older than John, and I earn twice as much money as John does." Nancy says, "I am one-half the age of John; John earns only one-third more money than I earn." John says, "I am happy with my life. My $90 000 house is worth 50% more than Dan's house; my $5200 sailboat is worth 30% more than Nancy's car." Dan replies, "My boat is worth three times as much as Nancy's car." Nancy replies, "I am happy earning $12 000 a year, and I have no debts."

    a. Enter as many values as you can in the following chart:

	Age	Salary	House	Boat	Car
Dan					
John					
Nancy					

    b. How much more valuable (in percent) is Dan's boat compared to John's boat?

Exercise Set 22.4, *Continued*

3. Janet, Sue, and Diana decide to compare the coins in their wallets. They each place their coins on a table. The following observations are made:

Diana has two pennies; Sue has twice as many pennies; Janet has triple Diana's number of pennies.

Sue has twice as many nickels as she has pennies; Diana has one-half as many quarters as she has pennies and one-half as many nickels as Sue.

Diana has as many dimes as Janet has pennies; Janet has one-half as many dimes as Diana; Sue has twice as many dimes as Janet has nickels.

Janet has as many quarters as Sue has pennies; Janet has one more nickel than she has quarters; Sue has as many quarters as Janet has dimes.

a. How many coins has each woman placed on the table?

b. What is the amount of money that each woman has placed on the table?

Prepare a chart.

Exercise Set 22.5

Time has an effect on age and on the distance traveled by a moving body. The next three problems involve the effects of time.

1. John, Jack, and Jim compare their ages. Today, Jim is twice as old as John. In ten years, Jack will be John's present age. Five years from now, John will be twenty.

   a. Make a chart that compares the ages of all three persons now, five years from now, and ten years from now.

   b. In ten years, what will be the age comparison of Jim to John in percent?

Exercise Set 22.5, *Continued*

2. A car is parked at a reference point. For two seconds, the driver accelerates the car at 4 m/s². The driver then stops accelerating and keeps the velocity constant for an additional four seconds.

   a. Complete the following chart.

Time	s	v	a
0			
1			
2			
3			
4			
5			
6			

   b. Construct the three graphs.

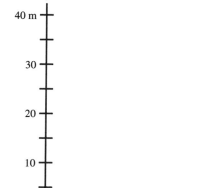

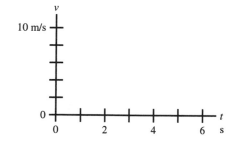

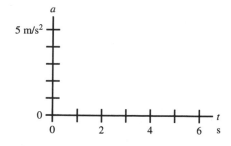

WORD PROBLEMS

3. Two aircraft leave Brisbane at the same time; both aircraft fly north.
Aircraft 1 flies at 500 mi/hr; aircraft 2 flies 50 mi/hr faster. At the same
time, a third aircraft leaves Smiles Island, which is 2000 miles north of
Brisbane. Aircraft 3 flies south at 450 mi/hr.

  a. Prepare a chart that indicates how far each aircraft is located to the
  north of Brisbane one, two, three, and four hours after takeoff.

  b. On the following graph, construct three lines that indicate the distance
  of each aircraft from Brisbane versus time:

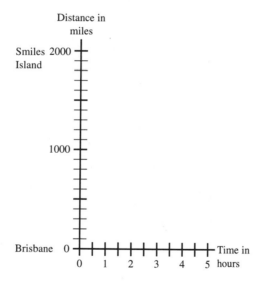

  c. Determine the (two) approximate times that aircrafts 1 and 2 will **meet**
  (intersect with) aircraft 3.

  d. When will aircraft 1 arrive at Smiles Island?

  e. When will aircraft 2 arrive at Smiles Island?

  f. When will aircraft 3 arrive at Brisbane?

Summary

Sheila Tobias discusses **Overcoming Math Anxiety** in her book of the same name. She notes that life experiences usually offer more than one path to take; there is seldom a wrong answer, and there is often more than one correct answer.

In most math books, there is usually one answer to a problem even though there may be more than one method for solving that problem. In this book, we have chosen and worded the problems so that there is only one correct answer or one set of correct answers.

In this chapter, we have given you a procedure for determining the solution to word problems. Students have successfully applied this procedure for at least three decades. Portions of this procedure may be used for simple word problems; all of this procedure should be used for complicated word problems.

Whether you realized it or not, we introduced very simple word problems starting with Chapter 1. Yet still, word problems may frighten you. Ms. Tobias describes these fears: The feelings you may experience are similar to the feelings an explorer may have felt when entering an unexplored jungle.

Observe our modern astronauts. They also feel anxious at times. They are thoroughly prepared prior to takeoff; yet unexpected events sometimes happen to them. They usually solve such unplanned problems and continue on with their mission. Therefore, you are normal if you are anxious at times when learning any new material such as math!

> *This chapter and the next have no chapter tests. You can consider each word problem to be a* ***mini-test.***

Wordy Word Problems

Excess Information

This chapter has been prepared to help you learn to examine the wording of a problem so that you can then extract the information that will be useful in solving the problem.

For most of the word problems in math texts, you are given only that information needed to solve the problem. That approach is not realistic. Problems in business, industry, and the home usually provide either too much information or too little information.

Each of the ten problems in the following Exercise Sets has been prepared to contain too much information. Each problem has much more information than you will need to answer the questions.

What should be your approach to such problems? You should read the problem once, determine the questions that are to be answered, and then extract only the information that will assist you in solving the problem. The space provided below each problem is provided so that you may document your problem solution on that page. If additional sheets are needed, label them with the exercise set number and the exercise number.

The six-part procedure of the previous chapter also applies to these wordy word problems. The only difference is that you must identify the information that you will need.

*As noted in the previous chapter, this chapter has no chapter test. Each word problem is a **mini-test.***

Exercise Set 23.1

The answers to the exercises in this chapter are given in Appendix A.

1. A neighborhood consists of 3 streets and 39 houses. Of these houses, 12 are single-family houses, 7 are two-family, 6 are three-family, and 4 are four-family apartments with two apartments on each floor. (The zoning code limits houses to two above-ground levels because of fire-truck ladder-height limitations.) A census is taken; the average number of persons per household in this neighborhood is 3.20.

    The number of houses and their colors are

    14 white,   8 green,   7 gray,   6 red,   3 blue,   1 a sickening violet

    The census taker has a computer printout that indicates there are a total of 38 people living in the one-family houses, 50 people living in the two-family houses, and 61 people living in the three-family houses. The other entry is not readable.

    a.   How many people live in the four-family apartments?

    b.   How many households are in this neighborhood?

    c.   What is the average number of people per household in each type of building structure?

    d.   The school district serving this neighborhood is overcrowded and splits the neighborhood in half. All of the houses, except for the single-family houses, are in the new school district. How many people live in the new school district?

Exercise Set 23.1, *Continued*

2. The president of an airline decides to study passenger **flying** habits at the airline's major terminal. The record of flights boarded indicates that

an average of 270 people fly on Sundays,

544	on Mondays,
322	on Tuesdays,
286	on Wednesdays,
310	on Thursdays,
650	on Fridays, and
189	on Saturdays.

The local bus line that serves the airport requests this information so that it can revise its total airport schedule. However, the airline managers are exploring special fares that should encourage midweek travel.

The local cab company that has the license to serve the airport wants to know the estimated effect of these special airline fares on taxi requirements and the number of bags of luggage each passenger might be carrying each day. Starting with Sunday, the airline luggage count is

521 on Sundays,

870 on Mondays,

708 on Tuesdays,

604 on Wednesdays,

732 on Thursdays,

1235 on Fridays, and

467 on Saturdays.

It is estimated that the special fare offerings will increase the Tuesday, Wednesday, and Thursday traffic by 20% each day. This will reduce the Monday and Friday seating requirements.

a. How many persons now travel each day of the week, in percent?

b. What is the average number of bags carried each week, and each day of the week, per person?

Exercise Set 23.2

1. A national consumer publication notes that consumers usually purchase one package of cheese at a time so that it will not become stale before being eaten. The typical consumer prefers to spend less than $2.00 on each package of cheese purchased.

   The prices of different brands of cheese are to be compared. The data from the labels on six packages are recorded by members of a consumer price-monitoring organization that monitors the local supermarkets weekly. The following chart is published in one of their weekly bulletins:

Cheese Brand	Net Weight	Package Shape	Selling Price	Package Covering
Jason's	12 oz	cube	$2.19	foil
Nathaniel's	8 oz	wedge	$1.39	paper
Lemieux's	252 g	cube	$1.69	paper
Best Buy	14 oz	wedge	$2.19	foil
Lauderleif	196 g	bar	$1.19	rag
Old Larder	10 oz	wedge	$1.49	foil

Conversion

1 oz = 28 g

Consumer Reports also plans a study to determine which is the least expensive per ounce: wrap, foil, rag, or paper.

a. What is the price per ounce for each brand of cheese?

b. Which brand of cheese is the best price per ounce of cheese?

c. Mrs. Smith is having a party. She is on a restricted budget but still wants to purchase a variety of cheeses for her party. What are the three lowest-price cheeses if she plans to purchase one bar, cube, or wedge of each?

Exercise Set 23.2, *Continued*

2. The following birth records are discovered in an old mining town:

   Alfredo and Marguarita marry; they have Jose, Sammy, and Alberto.

   Alberto and Barbara marry; they have Albert, Susan, and Maria.

   Benjamin and Dorris marry; they have William, Anita, Carlotta, Delores, and Ricardo.

   Jose and Anita marry; they have Carlos, Nancy, and Roberto.

   Pierre and Carlotta marry; they have Consuelo, Billy, Anatole, Jeanette, and Nellie.

   a. Who are the grandchildren of Alfredo and Marguarita?

   b. Who are the grandchildren of Benjamin and Dorris?

Exercise Set 23.3

1. There is to be a sale of surplus lumber. The sale will be limited to 1″-by-8″ boards that are 12 ft long. (The boards actually measure $\frac{3}{4}$″ by $7\frac{3}{4}$″ but are 12 ft long.)

    A family decides to construct one set of bookshelves using this lumber while it is on sale. This family gathers its books together and records the following information:

Person	Age	Tallest Book	Average Width	Number of Books	Weight of Books
Mom	39	12″	1.2″	30	72 lb
Dad	42	14″	1.6″	40	149 lb
Billy	14	11″	1.3″	30	80 lb
Nancy	11	12″	1.5″	34	104 lb
Norma	10	14″	0.8″	25	47 lb
Craig	8	10″	0.5″	16	14 lb

   a. How much shelf space will each person require?

   b. What are the average weight per book for each person and the average weight per book for the entire family?

   c. What is the average age of the persons in this household?

   d. Whose books are heaviest?

   e. Whose books will require the most space (width)?

   f. The family has a long hallway that can be used for the shelves. Therefore, the shelf length is not a problem. However, they want the height of the set of bookshelves to be as close to 4′6″ as possible, but no taller than 4′6″. If the bookshelves are to be constructed so that every shelf can contain the tallest book of any family member, then how many shelves can be constructed vertically?

Exercise Set 23.3, *Continued*

2. A small company must carefully schedule its employees. One small business has seven employees and one owner. The owner, Natalie, is salaried and earns $550 per week. This week Natalie worked 54 hours. The following information about the employees is available:

Jack prefers to play golf on weekends; he works 40 hours per week at $7.52 per hour.

Mary prefers a four-day week and works 6 hours each day as the company bookkeeper. Mary earns $6.84 per hour.

Al is the gardener and janitor. He works only when there is work to be done. This week it rained twice, so Al worked indoors 18 hours and outdoors 20 hours. Al is paid $5.66 per hour.

Jan is the chief assembler. She works a regular 40-hour week and contributes 6 extra hours of free overtime. She is paid $8.66 per hour.

Bob is the printed-circuit card pickler and is a specialist. This is a quiet week, so he works only 30 hours. He is paid $9.24 per hour.

Fred and Betty are the assemblers. They each work 40 hours per week and are paid $4.50 per hour.

a. What is the total payroll for this small company?

b. Because federal law requires that an additional 7% be added to the payroll for **Social Security,** what is the extra amount of payroll contribution the company must add to its payroll?

c. What are the average hours worked per person per week?

d. What are the average hours worked per person per day?

Exercise Set 23.4

1. A button inventory is being verified. The buttons that remain in eleven compartments are counted and identified as follows:

Quantity	Color	Diameter
66	red	5 cm
142	orange	5 cm
85	green	5 cm
210	blue	5 cm
66	purple	3 cm
93	black	5 cm
80	orange	4 cm
211	black	4 cm
96	red	3 cm
64	green	3 cm
99	black	3 cm

   a. What percent of the buttons are red, what percent are blue, and what percent are black?

   b. What percent of the buttons have a 4-cm diameter?

Twelve identical packages of buttons are prepared. Each contains six 5-cm green buttons, eight 4-cm black buttons, and four 3-cm red buttons.

   c. How many buttons remain in each of these three compartments?

   d. What percent of each of the eleven types of buttons now remain?

   e. Then, from the remaining inventory, how many packages can be assembled so that each package contains 5 red (3-cm) buttons, 3 green (3-cm) buttons, and 7 black (3-cm) buttons?

   f. How many of each of the three types of buttons selected will remain in the inventory?

Exercise Set 23.4, *Continued*

2. A recipe for providing a thickening agent requires that the following ingredients be prepared:

> Add 3 cups of flour and 2 tablespoons of salt to $\frac{1}{2}$ pound of water. Mix and stir thoroughly; warm on **medium** heat for 10 minutes.
>
> Let the mixture cool for 5 minutes. Then add 8 tablespoons of sugar and an additional 1 cup of water. Stir again, and heat this mixture until it is ready to boil.

If the pot is to be no more than one-half full, and the pots that are available are sized 1 quart, 2 quarts, 3 quarts, and 5 quarts, then which size pot should be chosen?

What is the ratio of dry to wet ingredients in the recipe?

Conversions

2 C = 1 pt = 32 T

16 T = 1 C

1 qt = 32 oz

2 pt = 1 qt

16 oz = 1 lb of water

"A pint's a pound,
the world around."
(Not any more!)

Exercise Set 23.5

1. A surveyor walks 40 feet away from a building. The building width is measured to be 72 ft 8 in. Each floor on this side of the building contains six windows.

   Two angles are measured by the surveyor: 46.32° from the ground to the top of the second floor, and 57.68° from the ground to the top of the third floor.

   a. What is the height of this three-story building?

   b. What is the distance from the top of the second floor to the top of this building?

Exercise Set 23.5, *Continued*

2. Three bank accounts are monitored by a private investigator during the months of June, July, and August. She prepares a chart that indicates the following deposits and withdrawals (to the nearest dollar), using a 24-hour automated **teller:**

Account Number	Starting Balance	June 2	July 4	August 13	Activity
			$1000	$3000	Deposit
1	$6427				
		$4000	$500	$1500	Withdrawal
		$2000	$1000		Deposit
2	$2158				
		$1000	$1000	$2000	Withdrawal
		$2000			Deposit
3	$3796				
			$2000	$1000	Withdrawal

The private investigator believes that money was withdrawn to pay for illegal bribes.

a. How much money was withdrawn during the months of June, July, and August?

b. How much money was deposited on the three indicated days?

c. Determine the difference between the total amount of money withdrawn and the total amount of money deposited.

A reminder: There is no chapter test for this chapter.
*Each exercise can be considered a **mini-test**.*

Numbers with Letters

Terms, Expressions, and Coefficients

Many math problems occur so often that we would like a "recipe" or "formula" for solving these problems over and over again. Study the following example.

Example At the end of every week, the owners of a small business add together all of the money to be paid to their workers. Then they subtract all of the deductions (such as taxes and insurances) and pay them to a branch of government or to an insurance company. The owners would find it helpful to devise a "recipe" or "formula" to use each week for their calculations.

Such a formula requires the use of letters to represent the various categories, which, in this case, are **time** and **money.** At the end of the week, the owners would replace those letters with the exact numbers (representing time and money) computed for each worker.

Formulas have been used for more than two thousand years. In 300 B.C., Aristotle used letters instead of numbers to identify the sides of triangles. A few years later, Euclid applied letters to geometry. In A.D. 800, traders brought the Hindu-Arabic numbers from India to the Middle East. At that time, accountants used letters as **accounting shorthand** for repetitive operations. The accountant's helpers had to memorize these **formulas.** They applied the formulas by inserting the appropriate numbers into the formulas and computing number (**numeric**) values.

When we use letters instead of numbers, we are working with that branch of math known as **algebra.**

Algebra is more general than arithmetic.

When we study algebra, we must distinguish between numbers and letters:

- The math name for **number** is **numeric.**
- The math name for **letter** is **literal.**

Refer to the definition of the math word **term** from Chapter 7 of Book 1. Sometimes a math expression is written that includes letters **(literals).** Such an expression is known as an **algebraic expression** whose terms contain numbers and letters. The meaning of the math word **term** does not change.

> In arithmetic, a **term** is a number (such as **5**), or a group of numbers (such as **12 × 2 ÷ 3**), whose math operations do not include addition or subtraction.

> In algebra, a **term** contains numbers (such as 5), letters (such as y), or both numbers and letters (such as $5y$) whose math operations do not include addition or subtraction.

Study the following examples of algebraic expressions that contain one or more terms. Each term is underlined.

Examples

$\underline{2ab}$	one term
$\underline{x} + \underline{y} - \underline{z}$	three terms
$\underline{3k} - \underline{d}$	two terms
$\underline{5jk} + \underline{3x} - \underline{4}$	three terms
$\underline{2x/y} - \underline{5xz}$	two terms
$\underline{7mr \div 3ms \cdot 6k}$	one term

The definition for **term** learned in Book 1 has been modified only to include letters as well as numbers. Otherwise, it has not changed. A term may contain a product such as $2ab$. A term may also include a quotient, such as $2w/y$.

In Book 1, we also noted that a **term** can be contained within grouping symbols. As an example, note the following expression. It is a one-term expression:

$$2 \cdot (4n + 5p - 6qr)$$

Why? Because the three (inner) terms within the parenthesis pair are considered to be one **grouped term.** Therefore, this expression is considered to be a one-term expression.

When numbers and letters are contained within one term, the number portion of the term is known as the **numeric coefficient,** or simply the **coefficient,** of the term. Examine the following term:

$$3mnp$$

The number **3** is the numeric coefficient of this single term.

Exercise Set 24.1

The answers to the exercises in this chapter are given in Appendix A.

Identify the term(s) within each expression. Then identify the coefficients of each term.

1. $3x - 4yz + 2xyz$

2. $5xy \div xz$

3. $7nmp \div q - 4nm + 9mp$

4. $abc \div k + 3bcd - 2ac + 5ad$

5. $4akz - 5mp/y + 5bdv$

Note that each **expression** contains one or more terms. When the expression contains literals as well as numerics, then the following math words are frequently used:

Expression	Math Word
one-term expression	**monomial**
two-term expression	**binomial**
three-term expression	**trinomial**
three-or-more-term expression	**polynomial**

These math words are used most frequently when algebraic fractions and factoring are being discussed or performed.

Like and Unlike Terms

Polynomial expressions contain two or more terms. These terms may be **like** or **unlike terms:**

> **Like terms** contain identical letters (or letters with exponents). An example of a polynomial expression containing three like terms is

$$3xy + 4xy - 5xy$$

> Each term has a single x and a single y and no other letters or exponents.

> **Unlike terms** contain letters different from the letters in all the other terms in that expression. An example of an expression containing unlike terms is

$$3xz + 4xy - 5yz$$

Why is it necessary to identify like terms? Because like terms in an expression can be combined, and thus, the expression is simplified.

Study the example on the following page.

Example   Simplify the following polynomial expression:

$$3xy - 2yz + 5xy$$

Answer   First identify the like terms; they are underlined.

$$= \underline{3xy} - 2yz + \underline{5xy}$$

If you find it helpful, group the like terms by rearranging the expression.

$$= (3xy + 5xy) - 2yz$$

Recall the definition of common factors from Chapter 8 of Book 1. Note that x and y are common factors. Factor xy.

$$= (3 + 5)xy - 2yz$$

Next combine the **3** and the **5.**

$$= 8xy - 2yz$$

We have reduced (**simplified**) this expression from three terms to two terms. Note that unlike terms cannot be combined.

In general, expressions should be reduced to the least possible number of terms. Therefore, all remaining terms will be unlike terms. Also recall the following from Book 1:

*The **value** of an expression must remain the same. Only the **form** of the expression may change.*

Exercise Set 24.2

Underline the terms in the following expressions. Then simplify these expressions by combining like terms.

1. $7xy + 3yz - 2xy$

2. $9ab - 2bc + 4ab$

3. $4mp - 7np + 3mn - 2np$

4. $5xyz + 3wxy - 2xyz + 4wxy$

5. $3abcd - 4bcd + 5 - 2abcd + 7bcd$

6. $5mnpq - 3mnpq + 7mnpq + 6mnpq$

7. $7rst - 5rt + 4rs - 3rt + 5rst$

8. $-9kmp + 8mp + kp - 2kmp - kp$

Hint: In the expressions in the following exercises, some terms contain letters that have not been arranged alphabetically. First, arrange all letters within a term in ascending alphabetical order.

9. $4xyz - 2wxy + 3zyx - 5xyw$

10. $6rs - 4ts + 11st + 2sr - 8rst$

The Concept of Equations

An **equation** is two equal expressions, known as **sides,** separated by an **equal (=)** sign. An equation contains at least one letter (literal). Separately, each side must follow the rule for an expression: The form may change but not the value. An example of an equation is

$$4x = 12$$

However, **both sides of the equation may be changed equally,** as described in the following list:

*Note: These **equal values** may be numbers or letters.*

- An equal value may be added to both sides.
- An equal value may be subtracted from both sides.
- An equal value may be multiplied into both sides if that value is not zero.
- An equal value may be divided into both sides if that value is not zero.
- The sides of an equation are equal. Thus, the sides can be reversed:

$$12 = 4x \quad \text{and} \quad 4x = 12$$

   are equivalent statements.

By custom, the term (or terms) containing the **unknown** appears on the left-hand side. Thus, the term (or terms) containing the known appears on the right-hand side. The object is to determine (or solve for) a value of the literal, such as x, that causes both sides of the equation to exactly balance, that is, to become equal.

Study the example on the following page.

Example   Examine the following equation and its operations:

$$7x - 3x = 16 - 4$$

Solve for x in this equation. In other words, determine the value of x that makes the equation become a true statement.

Answer   Combine like terms.

$$4x = 12$$

The unknown is x. Divide both sides by **4.**

$$4x \div 4 = 12 \div 4 \quad \text{and} \quad x = 3$$

Thus, the **solution** (or **answer, result, root**) to the equation

$$7x - 3x = 16 - 4$$

with x as the unknown is **3.**

How can I **verify (check)** that $x = 3$ is the correct solution? Substitute **3** for x in the original equation:

$$7(3) - 3(3) = 16 - 4 \quad \text{Is this true?}$$
$$21 - 9 = 16 - 4?$$
$$12 \equiv 12$$

Thus, $x = 3$ is the correct solution to the original equation.

Exercise Set 24.3

Determine the value of x for the following equations.

1. $6x = 72$

2. $8x = -48$

3. $-4x = 52$

4. $5x = 63 + 42$

5. $8x - 3x = 75$

6. $10x + 4x = 105 - 21$

7. $15x - 6x = 106 + 29$

8. $20x - 2x = 82 - 28$

The Four Basic Rules for Equations

When there is one unknown in an equation, it is necessary to "move" terms containing that unknown to one side (usually the left) of the equation. All of the terms containing knowns must be moved to the other side (usually the right) of the equation. As we have seen, both sides of an equation may be changed equally:

1. An equal value may be added to both sides.
2. An equal value may be subtracted from both sides.
3. An equal value may be multiplied into both sides if that value is not zero.
4. An equal value may be divided into both sides if that value is not zero.

*These are known as **the four basic rules for equations.***

Some equations have known terms and unknown terms on the same side. The four basic rules for equations continue to apply.

Move all terms containing the unknown to the left-hand side and all terms containing only knowns to the right-hand side. Now the "solution" for the unknown, such as x, may be determined.

Study the following example.

Example   Solve the following equation: $2x = 80 - 3x$.

Answer   Move $\mathbf{3x}$ to the left-hand side by adding $3x$ to both sides of the equation (because $-3x + 3x = 0$).

$$2x + 3x = 80 - 3x + 3x$$

The equation is now

$$5x = 80$$

Note that the term containing the unknown is on the left-hand side of the equation and the term containing the known is on the right-hand side of the equation.

Both sides may now be divided by 5 because $5x \div 5 = x$. The equation is now

$$x = 16$$

You may verify this result by substituting $x = 16$ into the original equation.

$$2(16) = 80 - 3(16) \quad \text{Is this true?}$$
$$32 = 80 - 48?$$
$$32 \equiv 32$$

We have "solved" the equation for x by first forcing all terms containing unknowns to the left-hand side and all terms containing knowns to the right-hand side, and then dividing both sides by the coefficient of the x-term. The result has been verified (or "checked").

The same technique is applied when known terms (that contain only numbers) are on the left-hand side, as the following example illustrates.

Example   Solve

$$6x - 5 = 7$$

Answer   We may "move" the **5** to the right-hand side by adding **5** to both sides of the equation.

$$6x - 5 + 5 = 7 + 5$$

The equation is now

$$6x = 12$$

Divide both sides by the coefficient **6** to determine x. The result is

$$x = 2$$

We have "solved" the equation for x by first forcing all terms containing knowns to the right-hand side and then dividing both sides by the coefficient of the x-term. Note that the single term with the unknown was already on the left-hand side.

Verify this result.

$$6(2) - 5 = 7 \quad \text{Is this true?}$$
$$12 - 5 = 7?$$
$$7 \equiv 7$$

Exercise Set 24.4

Solve for x in the following equations. Explain each step. Then verify your answer by substituting your result into the given original equation.

1. $7x = 12 + 4x$

2. $6x - 7 = 11$

3. $2x + 5 = 6x + 29$

4. $8x - 6 = 3x + 9$

Note that the answer to each of these exercises is an integer.

Again, solve for x in the following equations. Explain each step. Then verify your answer by substituting your result into the given original equation. The solution to each of these equations might **not** be an integer.

5. $3s + 8 = 5s - 10$

6. $5v - 17 = 9v + 12 - 6v$

7. $15 + 6y = 8 + 2y - 7$

8. $18t - 24 = 14t + 28$

9. $9Q - 44 = 16Q - 17$

As noted earlier, we "solve" an equation for its unknown, such as x, by first forcing all terms containing unknowns to the left-hand side and all terms containing knowns to the right-hand side. Then we divide both sides by the coefficient of the x-term. The unknown may be any letter. Typically, unknowns are chosen from the "end" of the alphabet. (This was suggested by René Descartes in the 1600s.) The following are typical unknowns:

$$p, \quad q, \quad r, \quad s, \quad t, \quad u, \quad v, \quad w, \quad x, \quad y, \quad z$$

When computer programming is involved, the unknown might consist of a group of letters or words, such as

PAY	AMOUNT	HOURS	TIME
PROFIT	LOSS	CASH	MILES

Each group of letters or words is considered to be one unknown.

Verifying Your Answer

Recall that the answer to an equation may be verified (checked) by inserting the calculated answer into the original equation. The **answer** is often referred to as the **result, root,** or **solution.** The value of each side must be the same number.

Study the following example.

Example   Solve the following equation:

$$7z - 5 = 4z + 10$$

Answer

$7z - 5 = 4z + 10$	Subtract $4z$ from both sides.
$7z - 5 - 4z = 4z - 4z + 10$	Combine like terms.
$3z - 5 = 10$	Add 5 to both sides.
$3z - 0 = 10 + 5$	Combine numeric terms.
$3z = 15$	Divide by 3.
$z = 5$	The solution for z is **5.**

Verify the result by substituting **5** into the original equation.

$$7(5) - 5 = 4(5) + 10?$$
$$35 - 5 = 20 + 10?$$
$$30 \equiv 30 \quad \text{The solution } (z = 5) \text{ is verified (or checked).}$$

Often the question mark (?) is shown above the equal sign.

$$\overset{?}{=} \quad \text{Such as } 35 - 5 \overset{?}{=} 20 + 10$$

This indicates that the equality has not yet been proven or demonstrated.

Exercise Set 24.5

Solve the following equations and verify (check) each result.

1. $17p - 18 = 24p - 4$

2. $85V + 72 = 54 + 49V$

3. $18z - 2z + 52 = 19 + 25z + 15$

4. $65 + 48Y - 15 = 55 - 88Y - 22$

5. $72 - 14u + 16 = 54u - 84 + 18u$

6. $17.6x - 14.2 = 3.5x + 28.1$

   *(The rules do **not** change for fractions.)*

7. $-71.2 - 44.4R - 82.7 = 26.5R + 59.3 - 17.6R$

8. $\dfrac{3}{2}x - 5 = 9x + \dfrac{5}{2}$

   (First multiply all terms of the equation by the least common denominator. See Chapter 9 of Book 1.)

Literal Equations

Equations that contain only letters are known as **literal** equations. An example of a literal equation is

$$s = vt \quad \text{(from Chapter 19)}$$

where s is the distance traveled by an object, v is the velocity of the object, and t is the amount of time that the object has been moving (traveling). Units are involved:

Let the units of s become meters and the units of t become seconds.

Thus, the units of v become meters per second.

This equation, in its present form, is most useful when v and t are given and s is to be computed. Consider the following example.

Example Determine the distance traveled if the velocity of a car is 40 meters/second north for 20 seconds.

Answer

$$s = vt$$
$$s = (40 \text{ meters/second north})(20 \text{ seconds})$$
$$= 800 \text{ meters north (or 800 meters scalar)}$$

Sometimes we desire to compute either the velocity of the object or the time during which the object is moving.

Example Determine the velocity of a train that has traveled a distance of 360 miles west in 5 hours.

Answer The equation $s = vt$ must be solved for v. One way is to reverse sides so that the desired unknown is on the left-hand side.

$$vt = s$$

Divide both sides by t.

$$v = s/t \quad \text{or} \quad v = s \div t$$

Substitute the known values for s and t.

$$v = (360 \text{ miles west}) \div (5 \text{ hours})$$
$$= 72 \text{ miles/hour west}$$

The sequence of the solution is not important. You may solve the equation for v as shown and then substitute the given values for s and t in the resulting equation. If you prefer, you may first substitute the values for s and t into the original equation ($vt = s$) and then solve for v.

$$v(5 \text{ hours}) = 360 \text{ miles west}$$
$$v(5 \text{ hours}) \div (5 \text{ hours}) = 360 \text{ miles west} \div (5 \text{ hours})$$
$$v = (360 \div 5) \text{ miles per hour west}$$
$$v = 72 \text{ miles per hour west}$$

Exercise Set 24.6

Solve the following problems and verify (check) each result.

1. Determine the velocity of a truck that travels a distance of 400 miles east in 8 hours. Show all your steps, starting with $s = vt$.

2. Determine the time that a runner requires to travel a distance of 500 meters at a speed (a scalar) of 4 meters per second. Show all your steps, starting with $s = vt$.

3. Determine the height h of a rectangle whose width w is 70 meters and whose area A is 420 square meters. Show all your steps, starting with $A = hw$.

4. Determine the length (in meters) of the side s of a square whose area A is 6400 square meters. Show all your steps, starting with $A = s^2$.

Summary

Expressions may involve both numbers (**numerics**) and letters (**literals**). As presented in Book 1, an expression may change in form but not in value. The presence of letters does not affect this requirement.

When numbers and letters are contained in one term, the number portion of the term is the **numeric coefficient.** Examine the term $3mnp$. The number **3** is the numeric coefficient of this single term.

A **polynomial** is an expression that consists of many terms. The five-term expression

$$3xy + 5yz - 2xz + 7x - 4y$$

is an example.

Like terms can be combined. **Unlike terms** cannot be combined until one (or more) of the letters is replaced with a number.

Equations consist of two expressions that are separated by an equal ($=$) sign. (Also, equations must have at least one literal.)

Four laws govern equations:

1. An equal value may be added to both sides of an equation.
2. An equal value may be subtracted from both sides of an equation.
3. An equal value may be multiplied into both sides of an equation if that value is not zero.
4. An equal value may be divided into both sides of an equation if that value is not zero.

These equal values may be either numbers or letters.

We "solve" an equation for its unknown, such as x, by first forcing all terms containing unknowns to the left-hand side, next forcing all terms containing knowns to the right-hand side, and then dividing both sides by the coefficient of the x-term. The unknown may be any letter. That letter is usually chosen from the "end" of the alphabet.

An example follows on the next page.

Example   Solve the following equation:

$$7x - 5 = 2x - 25$$

Answer

$7x - 5 = 2x - 25$	Subtract 2x from both sides.
$7x - 2x - 5 = 2x - 2x - 25$	Combine like terms.
$5x - 5 = -25$	Add **5** to both sides.
$5x - 5 + 5 = -25 + 5$	Combine like terms.
$5x = -20$	Divide both sides by **5.**
$x = -4$	The solution is **−4.**

To verify the solution, substitute -4 into the original equation.

$7(-4) - 5 = 2(-4) - 25?$	Remove parentheses.
$-28 - 5 = -8 - 25?$	Combine terms.
$-33 \equiv -33$	The result is verified.

The symbol $\overset{?}{=}$ may be used if desired. (It replaces the ? at the end of each equation and indicates that the equality is not yet verified.)

Literal equations (equations that contain only letters) may be solved for any one of the unknown letters. Numbers may then be substituted for the remaining letters. For example,

$$s = vt$$

may be solved for *v.* The resulting equation is

$$v = s \div t$$

Values for *s* and *t* may now be substituted and the corresponding value of *v* determined.

Glossary of Math Words

Algebra (al′-je-bruh)   Expressions and equations that contain signs, numbers, and letters; equations also contain one **equal (=)** symbol.

Answer (an′-ser)   One name given to the solution to an equation; also known as the **result, root,** and **solution.**

Binomial (by-no′-mee-ul)   A two-term expression.

Check (check′)   To determine if the solution to an equation is correct. *See* Verify.

Coefficient (co-uh-fish′-ent)   The numeric portion of an algebraic term.

Equation (ee-kway′-shun)   Two expressions separated by an equal (=) sign.

Like (lyke) **terms**   Terms whose letters (and exponents) are the same, such as $4x^2y$ and $-9x^2y$.

Literal (lit′-ur-ul)   The math name for "letter."

Monomial (mo-no'-mee-ul)   A one-term expression.

Numeric (new-mair'-ik)   The math name for "number."

Polynomial (pol'-ih-no'-mee-ul)   A many-term expression.

Result (ree-zult')   One name given to the solution to an equation; also known as **answer, root,** and **solution.**

Root (root')   One name given to the solution to an equation; also known as **answer, result,** and **solution.**

Solution (so-loo'-shun)   One name given to the answer to an equation; also known as **answer, result,** and **root.**

Term (turm')   In algebra, the product (and perhaps also the quotient) of a sign, number, and letter(s).

Trinomial (try-no'-mee-ul)   A three-term expression.

Unknown (un-nohn')   In an equation, the letter, or group of letters, whose value is to be determined; typical unknown letters are *p, q, r, s, t, u, v, w, x, y,* and *z.*

Unlike (un'-lyke) **terms**   Terms whose letters and exponents are different, such as $4x^2y$ and $-9xy^2$.

Verify (ver'-uh-fy)   To determine if the solution to an equation is correct. *See* Check.

Chapter 24 Test

Follow instructions carefully:

> *On a separate piece of paper, write the answers to the following questions. Do* not *write on these pages.*
>
> *When you are finished, compare your answers with those given in Appendix B.*
>
> *Record the date, your test time, and your score on the chart at the end of this test.*

Simplify the following expressions by combining like terms.

1. $8mn - 13mn + 2mn$

2. $7pqr + 9qrs - 3pqr$

3. $18xy + 51xyz - 23xy + 17xyz$

4. $12bc + 31ab - 15 + 4bc$

5. $4rs - 15rst - 18sr - 9tsr$

Solve for the value of the unknown; verify that solution.

6. $5x = -15$

7. $7y = 31 - 3$

8. $+6z - 8z = -22$

9. $9t - 13t = 8 - 18$

10. $13k = 12 + 7k$

11. $-21p - 12 = 16$

12. $17 + 31v = 86v - 5$

13. $32 - 4d = -21d + 11 + 16d$

14. $16.2g - 13.18 = 31.76g + 28.054$

15. $11h + 17/3 = 3h - 7/3 - 4h/3$

Rewrite each equation so that the desired unknown letter or word is on the left-hand side. Then substitute the given values for the other unknown letters, and determine the desired numeric value. Include units in your answer.

16. $p = 4s$ Solve for s where $p = 125.2$ cm.

17. $s = vt$ Solve for v where $t = 5$ sec and s $= 65$ ft south.

18. $P = 2L + 2W$ Solve for L where $P = 1.45$ ft and $W = 2.6$ in.

19. $\cos \theta = $ adjacent $\div$ hypotenuse   Solve for the hypotenuse where the adjacent side is 117 ft and θ is 42°.

20. $a_n = a_1 + (n - 1)d$ Solve for d where $a_1 = 4$ and $a_5 = 324$.

Chapter 24 Test Record

DATE	TIME	SCORE

Refer to

- Appendix B for the correct answers to this test.
- Appendix C if taking this test required too much effort.
- Appendix D for **Memory Methods** assistance.
- Appendix E if your test scores are decreasing.

When a problem seems difficult, find one like it in this chapter. Then study that (and the related material) again.

Develop additional 3 × 5 cards for those ideas, problems, and procedures that caused you difficulty.

Part 8 Review Test

CHAPTER 22
Word Problems

CHAPTER 23
Wordy Word Problems

CHAPTER 24
Numbers with Letters

Instructions

Review your 3 × 5 cards before you take this test.

Complete this test.

You may find some ideas, problems, or procedures difficult or confusing. Develop 3 × 5 cards for them.

Part 8 Review Test

A bicyclist is traveling in Nova Scotia on a vacation. She leaves Halifax and travels North toward Pugwash, a distance of 120 km. After 3 hours, she realizes that she has traveled 90 km.

1. What is her average velocity?

2. How many hours will it take her to reach Pugwash from Halifax?

3. Prepare a graph of her trip from Halifax to Pugwash with distance traveled (in km) on the *y*-axis versus time (in h) on the *x*-axis.

4. Maria, Jose, and Carmen compare their ages. Carmen notes that she is 7 years older than Maria. Maria replies that she is half the age of Jose. Given that Jose is 42 years old, determine the ages of Maria and Carmen. Record your answers in the chart of Problem 5.

5. The same day, Jose, Carmen, and Maria compare their salaries. Jose remarks that he earns 21% more than Carmen. Maria notes that she earns 33% less than Jose. It is known that Carmen earns $26 000 per year. How much do Jose and Maria earn? Record your answers in the following chart:

Person	Salary per Year	Age in Years
Jose		
Carmen		
Maria		

A rectangular table top is to be painted. The length of one of its sides is 6.5 ft, and the length of another side is 1 yard shorter. Two circles are cut in the table top. Circle A has a diameter of 3 in.; circle B has a circumference of 8 cm.

6. What is the area of the table top before the circles are cut into it?

7. What is the area of circle A?

8. What is the area of circle B?

9. How much table top remains to be painted (after the circles have been cut out of it)?

The fourth term of a geometric progression is 625, and the third term is 125.

10. Determine the value of the first term.

11. Determine the value of the second term.

12. Determine the value of the four-term sequence (known as the **series**).

13. Determine the value of the fifth term.

A runner can run 400 meters in 0.96 minute.

14. What is her speed in feet per second?

15. How many miles can she run in 3 minutes?

Simplify the following expressions by combining like terms.

16. $6a + 5ab - 5a$

17. $8x + xy - 3yx + x$

Solve each equation for the unknown letter.

18. $11y = 8y - 9$

19. $-8 + 21z = -9z - 20$

20. $s = vt$ Solve for t where

$$s = 64 \text{ meters north}$$
$$v = 4 \text{ meters/second north}$$

Part 8 Review Test Record

DATE	TIME	SCORE

Refer to

- Appendix B for the correct answers to this test.
- Appendix C if taking this test required too much effort.
- Appendix D for **Memory Methods** assistance.
- Appendix E if your test scores are decreasing.

When a problem seems difficult, refer to the pages references in Appendix B. Then study that (and the related material) again.

Develop additional 3 × 5 cards for those ideas, problems, and procedures that caused you difficulty.

Book 2 Test

1. Simplify and reduce the following fraction:

$$\frac{63/5}{9/5}$$

2. Simplify and reduce the following fraction:

$$\frac{\dfrac{9}{2} \div \dfrac{7}{8}}{\dfrac{18}{25} \div \dfrac{4}{5}}$$

3. Simplify and reduce the following fraction:

$$\frac{\dfrac{5}{6} - \dfrac{4}{9}}{\dfrac{4}{5} + \dfrac{7}{15}}$$

4. What is the decimal fraction for 5/8?

5. Multiply 32.02 by 6.9.

6. Subtract 3.74 from 209.1.

7. Convert the following expression to exponential notation:

$$3 \times 5 \times 3 \times 3 \times 5 \times 3 \times 5$$

Calculate the numbers represented by the following expressions.

8. $(4^2)^{-1}$

9. $(3^4 \div 3^5)^2$

10. Write the following number using scientific notation:

0.000 7

11. Determine the numeric value for the following expression:

$$7 \times 10^5$$

Evaluate the following expressions.

12. $\sqrt[3]{64}$

13. $(81)^{1/4}$

14. A square has a side that is 6.4 in. long. Determine its perimeter.

15. A rectangle measures 3.8 cm on one side and 4.6 cm on the other side. Determine its area.

16. The diameter of a circle is 4.51 cm. Determine its area.

A cube is 21.8 in. on one side.

17. Determine its volume.

18. Determine its surface area.

19. A sphere has a radius of 5.3 in. Determine its volume.

There are 2 households on a short street. The ages of the 5 persons in the first household are

$$18, \quad 19, \quad 44, \quad 65, \quad 69$$

The average age of the 4 persons in the second household is 28.

20. What is the average age of the persons in the first household?

21. What is the average age of the persons in the 2 households combined?

A street contains 11 houses. The prices of the 11 houses are

$225\ 000, \quad \$180\ 000, \quad \$210\ 000, \quad \$225\ 000, \quad \$305\ 000,$
$275\ 000, \quad \$210\ 000, \quad \$230\ 000, \quad \$275\ 000, \quad \$190\ 000, \quad \$225\ 000$

22. Determine the mean average price.

23. Determine the mode price, and display this information on a bar chart.

24. Determine the percent of houses that are priced at $225\ 000$ or less.

25. Determine the slope of a line constructed between point $A(-8, -5)$ and point $B(-1, -5)$.

26. Determine the slope of a line constructed between point $A(-6, 7)$ and point $B(3, 2)$.

27. Determine the slant line distance between points *A* and *B* in Problem 26.

Two bicyclists are in Nova Scotia, one in Halifax and the other in Pugwash. The first bicyclist leaves Halifax and travels at a velocity of 25 km/h north toward Pugwash. The second bicyclist leaves Pugwash at the same time and travels south toward Halifax 5 km/h faster than the first bicyclist. (It is known that Pugwash is 120 km north of Halifax.)

28. What is the distance of each bicyclist from Halifax at the end of 0, 1, 2, 3, and 4 hours?

29. Sketch the trip of each bicyclist (two lines) on the following graph:

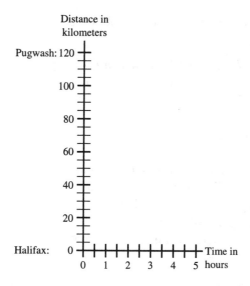

30. From the preceding graph, determine approximately where the two bicyclists will meet.

31. Approximately when will the first bicyclist reach her destination?

32. Approximately when will the second bicyclist reach his destination?

Solve the following problems.

33. What is the fourteenth even integer?

34. What is the sum of all the even integers from **2** to the twenty-third integer?

35. The third term of an arithmetic progression is 19; the fourth term is 24. Determine the value of the first term.

36. The fourth term of a geometric progression is 189, and the third term is 63. Determine the value of the first term.

37. Convert a $7\pi/4$-radian angle to degree measure.

38. In what quadrant is the vector of Problem 37 located?

39. Convert a $-590°$ angle to radian measure.

40. In what quadrant is the vector of Problem 39 located?

41. A person wants to determine the height of a building. She walks exactly 150 ft from the corner of this building. The angle from the horizontal to the top of the building is measured to be 61.5°. What is the height of this building?

A student leaves work at 4:30 P.M. and travels 30 miles from her job to her college campus. Her estimated time of arrival (ETA) is 6:00 P.M.

There are 5280 ft in 1 mi, 60 min in 1 hr, and 60 sec in 1 min.

42. What is her average speed in miles per hour?

43. What is her average speed in feet per second?

Three friends, Jane, Alice, and Karl, compare their ages and salaries. Jane says, "I am 29 years old, and I earn $24 000 per year." Alice states, "You are 7 years older than I am, and I earn 25% more than you do." Karl quickly notes, "Well, I am only 3 years younger than you, Alice, and I earn 41% less than you earn." (Perhaps it is not always wise for friends to compare ages and salaries!)

44. Complete the following chart by determining each person's age and salary:

Person	Age in Years	Salary in Dollars/Year
Jane		
Alice		
Karl		

Maria, Jose, and Carmen compare their ages. In 5 years, Jose will be twice (two times) as old as Maria is at that time. In 10 years, Carmen will be 2 years younger than Maria is now. Jose's present age is 55.

45. Complete the following chart by determining each person's age now, 5 years from now, and 10 years from now:

Person	Age Now (Present Age)	Age in 5 Years	Age in 10 Years
Maria			
Jose			
Carmen			

46. Simplify the following expression:

$$13xy + 29xz - 5 + 14xy$$

Solve each equation for the unknown letter.

47. $18p - 26p = 16 - 36$

48. $7 + 12t = 32t - 5$

49. $y = mx + b$ where $m = 2$, $x = 5$, and $b = 3$

50. $v = at$ where $v = 24$ meters/second west and
$a = 4$ meters/second^2 west

Book 2 Test Record

DATE	TIME	SCORE

Refer to

- Appendix B, for the correct answers to this test.
- Appendix C if taking this test required too much effort.
- Appendix D for **Memory Methods** assistance.
- Appendix E if your test scores are decreasing.

When a problem seems difficult, locate the pages referenced in Appendix B. Then study that (and the related material) again.

Develop additional 3 × 5 cards for those ideas, problems, and procedures that caused you difficulty.

A Note from the Authors

These math books have been devised to explain arithmetic operations to you and to prepare you for that branch of mathematics known as **algebra** and for other technical or business subjects.

You should have solved all (or most) of the problems in this book, and you should have tested yourself at the end of each chapter. Whenever you discovered that you had forgotten a method, procedure, notation, or rule, then you should have reviewed the material and retested yourself.

> IF you have worked and reworked our examples, THEN you are well prepared to study more math, technical, or business subjects.

> IF you have some math anxiety with these subjects, THEN review the appropriate chapter(s) of these books.

Most of us have some anxiety when starting on a new journey. This may involve moving to a new place, starting a new job, or beginning a new math course.

In these books, we have explained each new math concept, described its source where known, and then had you practice, practice, practice. We want you to develop both new skills and confidence in your abilities.

We have explored most of the fundamentals of algebra, geometry, trigonometry, calculus, and statistics in these books using only numbers or numbers and formulas. We have purposely avoided using letters whenever numbers could explain the fundamentals. We firmly believe that most math concepts can be explored using numbers.

We congratulate you for your persistence in completing these two books. (But be sure to continue on with the Appendices!) You must, by now, agree with Euclid's statement to King Ptolemy:

> There is no royal road to mathematics.

During the writing of these books, we were encouraged by the comments of math teachers. Students with math anxieties have noted that these books should be very helpful to other people who have a wide variety of math-related difficulties.

We sincerely hope you have found our books helpful. You have our best wishes!

Bob Angus
Claudia Clark

Answers to Exercise Sets

CHAPTER 15

Exercise Set 15.1

1. $(1/5) \times (4/3) = 4/15$
2. $(3/8) \times (4/1) = 3/2$
3. $(1/6) \times (4/3) = 2/9$
4. $(5/12) \times (18/5) = 3/2$
5. $(5/12) \times (18/1) = 15/2$
6. $(5/18) \times (24/7) = 20/21$

Exercise Set 15.2

1. $(1/2) \times (3/4) \times (8/3) \times (6/1) = 6$
2. $(1/2) \times (3/4) \times (8/3) \times (1/6) = 1/6$
3. $(1/2) \times (4/3) \times (8/3) \times (1/6) = 8/27$
4. $(1/2) \times (3/4) \times (8/3) \times (1/5) = 1/5$

Exercise Set 15.3

1. LCD for all four fractions: $5 \times 3 \times 2$
   $(18 - 10) \div (25 + 15) = 1/5$
2. LCD for all four fractions: $3 \times 3 \times 5$
   $(20 + 18) \div (6 - 15) = -38/9$
3. LCD for all four fractions: $2 \times 3 \times 3$
   $(3 + 4) \div (12 + 8) = 7/20$
4. LCD for all four fractions: $2 \times 7 \times 3$
   $(3 + 18) \div (8 - 7) = 21/1 = 21$
5. LCD for all four fractions: $2 \times 2 \times 5$
   $(15 - 16) \div (16 - 10) = -1/6$
6. Denominator LCD: 5×2
   $[(3/4) \times (4/5)] \div [(4/5) - (1/2)] = [(3/4) \times (4/5)] \div [8/10 - 5/10]$
   $= (3/4) \times (4/5) \times (10/3) = 2$

CHAPTER 16

Exercise Set 16.1

1. $3 \times 3 \times 3 \times 3 \times 3 \times 3 = 729$
2. $4 \times 4 = 16$
3. $10 \times 10 \times 10 = 1000$
4. $2 \times 2 \times 2 \times 2 \times 2 \times 2 \times 2 \times 2 \times 2 = 512$
5. $8 \times 8 \times 8 = 512$
6. $1 \times 1 \times 1 \times 1 \times 1 \times 1 \times 1 \times 1 \times 1 = 1$
7. $5 \times 5 \times 5 \times 5 = 625$
8. $25 \times 25 = 625$
9. $2^1 = 2$

Exercise Set 16.2

1. 5 is the base; 4 is the power; $5 \times 5 \times 5 \times 5 = 625$.
2. 21 is the base; 2 is the power; $21 \times 21 = 441$.
3. 2 is the base; 10 is the power; $2^{10} = 1024$.
4. 10 is the base; 2 is the power; $10 \times 10 = 100$.
5. 7 is the base; 3 is the power; $7 \times 7 \times 7 = 343$.
6. 3 is the base; 3 is the power; $3 \times 3 \times 3 = 27$.
7. 7 is the base; 5 is the power; $7 \times 7 \times 7 \times 7 \times 7 = 16\ 807$.
8. 12 is the base; 3 is the power; $12 \times 12 \times 12 = 1728$.
9. 2 is the base; 20 is the power; $2^{20} = 1\ 048\ 576$.
10. 10 is the base; 4 is the power; $10 \times 10 \times 10 \times 10 = 10\ 000$.
11. 6 is the base; 4 is the power; $6 \times 6 \times 6 \times 6 = 1296$.
12. 3 is the base; 5 is the power; $3 \times 3 \times 3 \times 3 \times 3 = 243$.

Exercise Set 16.3

1. $3^{2+3} = 3^5 = 243$
2. $5^{3+2} = 5^5 = 3125$
3. $7^{1+2} = 7^3 = 343$
4. $2^{4+6} = 2^{10} = 1024$
5. $1^{3+5} = 1^8 = 1$
6. $4^{2+3} = 4^5 = 1024$
7. $7^{3+2} = 7^5 = 16\ 807$
8. $3^{1+2} = 3^3 = 27$
9. $2^{6+7} = 2^{13} = 8192$
10. $2^3 \times 1^7 = 8 \times 1 = 8$

Exercise Set 16.4

1. $3^{7-4} = 3^3 = 27$
2. $4^{7-6} = 4^1 = 4$
3. $5^{8-6} = 5^2 = 25$
4. $1^{9-6} = 1^3 = 1$
5. $5^{7-5} = 5^2 = 25$
6. $6^{9-4} = 6^5 = 7776$
7. $9^{8-6} = 9^2 = 81$
8. $7^{6-4} = 7^2 = 49$

Exercise Set 16.5

1. $9^0 = 1$
2. $(96)^0 = 1$
3. $(86)^0 = 1$
4. $1^0 = 1$
5. $(27)^0 = 1$
6. $(-19)^0 = 1$
7. $(2\ 037\ 605)^0 = 1$
8. $(-1/26)^0 = 1$

Exercise Set 16.6

1. $2^{-2} = 1/2^2$
2. $2^2 = (1/2)^{-2}$
3. $1/2^2 = 2^{-2}$
4. $5^{-4} = 1/5^4$
5. $7^5 - 1/7^{-5}$
6. $1/9 = 1/9^1 = 9^{-1}$
7. $1 \times 10^8 = 10^8 = 1/10^{-8}$
8. $1 \times 10^{-6} = 10^{-6} = 1/10^6$
9. $1/4 = 1/4^1 = 4^{-1}$
10. $4^7 = 1/4^{-7}$

Exercise Set 16.7

1. $2^{-3} = 1/2^3 = 1/8 = 0.125$
2. $3^{-2} = 1/3^2 = 1/9 = 0.111...$
3. $5^{2-5} = 5^{-3} = 1/5^3 = 1/125 = 0.008$
4. $6^{-2} = 1/6^2 = 1/36 = 0.027...$
5. $7^{5-6} = 7^{-1} = 1/7 = 0.\overline{142857}$
6. $8^{-1} = 1/8 = 0.125$
7. $2^{-2} = 1/2^2 = 1/4 = 0.25$
8. $4^{-3} = 1/64 = 0.015625$

Exercise Set 16.8

1. $3^{2\times3} = 3^6 = 729$
2. $2^{2\times3} = 2^6 = 64$
3. $4^{2\times3} = 4^6 = 4096$
4. $5^{1\times2} = 5^2 = 25$
5. $2^{-2\times3} = 2^{-6} = 1/2^6 = 1/64 = 0.015625$
6. $2^{(-2)\times(-3)} = 2^6 = 64$
7. $3^{(-3)\times(-3)} = 3^9 = 19\ 683$
8. $7^{(-2)\times4} = 7^{-8} = 1/7^8 = 15\ 764\ 801$
9. $1^{(-2)\times5} = 1^{-10} = 1/1^{10} = 1$
10. $2^{(-2)\times0} = 2^0 = 1$
11. $10^{(-2)\times3} = 10^{-6} = 1/10^6 = 1/1\ 000\ 000 = 0.000\ 001$
12. $10^{(-2)\times(-3)} = 10^6 = 1\ 000\ 000$

Exercise Set 16.9

1. $2^6 \times 4^6 = 64 \times 4096 = 262\ 144$
2. $3^6 \times 2^6 = 729 \times 64 = 46\ 656$
3. $2^9 \times 3^6 = 512 \times 729 = 373\ 248$
4. $5^{12} \times 2^{12} = 1 \times 10^{12}$
5. $2^4 \times 4^6 = 16 \times 4096 = 65\ 536$
6. $3^6 \times 2^{-6} = 729 \times 1/64 = 11.390\ 625$
7. $2^4 \times 4^{-6} = 16 \times 1/4096 = 1/256 = 0.003\ 906\ 25$
8. $3^4 \times 4^{-6} = 81 \times 1/4096 = 0.019\ 775\ 4$
9. $2^6 \times 4^9 = 64 \times 262\ 144 = 16\ 777\ 216$
10. $3^4 \times 5^{-6} = 81 \times 1/15\ 625 = 0.005\ 184$
11. $4^{15} \times 4^{-15} = 4^0 = 1$
12. $5^6 \times 2^{10} = 15\ 625 \times 1024 = 16\ 000\ 000$
13. $2^{12} \div 5^{12} = 4096 \div (24\ 414 \times 10^8) = 0.000\ 016\ 8$
14. $2^{12} \div 2^8 = 2^4 = 16$
15. $3^4 \div 3^8 = 3^{-4} = 1/3^4 = 0.012\ 345\ 7$
16. $5^6 \div 5^{10} = 1/5^4 = 0.001\ 6$
17. $6^6 \div 2^4 = 46\ 656 \div 16 = 2916$
18. $2^{12} \div 5^{-12} = 2^{12} \times 5^{12} = 1 \times 10^{12}$
19. $3^{12} \div 2^{-8} = 3^{12} \times 2^8 = 1.360\ 5 \times 10^8$
20. $3^4 \div 3^8 = 3^{4-8} = 3^{-4} = 0.012\ 345\ 7$
21. $5^6 \div 5^{-10} = 5^{16} = 1.525\ 9 \times 10^{11}$
22. $6^{-6} \div 2^4 = 1/6^6 \times 1/2^4 = 1/746\ 496 = 0.000\ 001\ 339\ 591\ 9$

Exercise Set 16.10

1. $8 - 4 = 4$
2. $4 - 8 = -4$
3. $9 + 8 = 17$
4. $9 - 8 = 1$
5. $9 + 8 - 16 = 1$
6. $1/8 + 8 = 8.125$
7. $9 + 27 = 36$
8. $81 + 1/4 - 8 = 73.25$
9. $25 - 32 = -7$
10. $1/16 + 16 - 1/25 = 16.022\ 5$

Exercise Set 16.11

1. 2×10^{-5}
2. 4.72×10^3
3. 3.5×10^{-2}
4. 6.03×10^4
5. 2.7×10^{-7}
6. 2.45×10^8
7. -2.7×10^4
8. 3.06×10^{-4}
9. -8.29×10^{-2}
10. 5.2728×10^4
11. 9.2716×10^{-5}

Exercise Set 16.12

1. 5000
2. $7.8 \times 10^{-5} = 0.000\ 078$
3. $-3.92 \times 10^8 = -392\ 000\ 000$
4. $9.267 \times 10^8 = 926\ 700\ 000$
5. $-7.825 \times 10^{10} = -78\ 250\ 000\ 000$
6. $-9.2866 \times 10^0 = -9.2866$
7. 1932
8. 0.000 827 48
9. -315.19
10. $-0.000\ 097\ 25$
11. 84 200 000
12. 0.000 027 748

Exercise Set 16.13

1. $(2.1 \times 4) \times 10^{(2+3)} = 8.4 \times 10^5$
2. $(8.4 \div 2) \times 10^{(5-3)} = 4.2 \times 10^2$
3. $(2.6 \times 3) \times 10^{(5-2)} = 7.8 \times 10^3$
4. $(5.7 \div 3) \times 10^{4-(-2)} = 1.9 \times 10^6$
5. $(6 \times 4 \div 8) \times 10^{(5+2)-3} = 3 \times 10^4$
6. $[(3.6 \times 5) \div (3 \times 6)] \times 10^{(5+2)-[4+(-3)]} = (18 \div 18) \times 10^{(7-1)} = 1 \times 10^6$
7. $60 \times 10^2 \div 6 \times 10^3 = 10 \times 10^{-1} = 1 = 1 \times 10^0$

Exercise Set 16.14

1. 26×10^{-3} grams = 26 milligrams
2. 37×10^3 meters = 37 kilometers
3. 28.2×10^3 grams = 28.2 kilograms
4. 50×10^{-3} s = 50 milliseconds
5. 27×10^3 liters = 27 kiloliters
6. 40×10^{-3} liters = 40 milliliters
7. 40×10^3 meters/hour = 40 kilometers/hour
8. 6×10^3 kilobytes = 6 megabytes
9. 5×10^3 megabytes = 5 gigabytes
10. 186×10^3 miles/hour = 186 kilomiles/hour
11. 276×10^{-3} milligrams = 276 micrograms
12. 3.548×10^{-3} microseconds = 3.548 nanoseconds
13. 26×10^{-3} grams = 26 mg
14. 37×10^3 meters = 37 km
15. 28.2×10^3 grams = 28.2 kg

Exercise Set 16.14, *Cont.*

16. 50×10^{-3} sec = 50 ms
17. 27×10^{3} liters = 27 kL
18. 40×10^{-3} liters = 40 mL
19. 40×10^{3} meters/hour = 40 km/h
20. 6×10^{3} kilobytes = 6 Mbytes
21. 5×10^{3} Mbytes = 5 Gbytes
22. 186×10^{3} mi/hr = 186 kmi/hr
23. 276×10^{-6} mg = 276 ng
24. 3.548×10^{-3} μs = 3.548 ns

Exercise Set 16.15

1. $(25)^{1/2} = (5 \times 5)^{1/2} = 5$
2. $(16)^{1/2} = (4 \times 4)^{1/2} = 4$
3. $+8$ and -8
4. $(27)^{1/3} = (3 \times 3 \times 3)^{1/3} = 3$
5. $(216)^{1/3} = (6 \times 6 \times 6)^{1/3} = 6$
6. $(512)^{1/3} = (8 \times 8 \times 8)^{1/3} = 8$
7. $7^2 = 7 \times 7 = 49$
8. $7^3 = 7 \times 7 \times 7 = 343$
9. $(-2)^2 = (-2) \times (-2) = 4$
10. $(-2)^3 = (-2) \times (-2) \times (-2) = -8$
11. $-(2)^2 = -(2) \times (2) = -4$

Exercise Set 16.16

1. $(6561)^{1/2} = (3^8)^{1/2} = 3^{8/2} = 3^4 = 81$
2. $(3375)^{1/3} = (5^3 \times 3^3)^{1/3} = 5^{3/3} \times 3^{3/3} = 5^1 \times 3^1 = 15$
3. $(256)^{1/4} = (2^8)^{1/4} = 2^{8/4} = 2^2 = 4$
4. $(169)^{1/2} = (13^2)^{1/2} = 13^{2/2} = 13$
5. $(243)^{1/5} = (3^5)^{1/5} = 3^{5/5} = 3^1 = 3$
6. $+25$ and -25

CHAPTER 17

Exercise Set 17.1

1. $P = 4 \times (7 \text{ cm}) = 28$ cm
   $A = (7 \text{ cm}) \times (7 \text{ cm}) = 49 \text{ cm}^2$

Exercise Set 17.1, *Cont.*

2. $P = 4 \times (12 \text{ in.}) = 48 \text{ in.}$
$A = (12 \text{ in.}) \times (12 \text{ in.}) = 144 \text{ in.}^2$

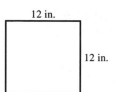

3. $P = 4 \times (8 \text{ cm}) = 32 \text{ cm}$
$A = (8 \text{ cm}) \times (8 \text{ cm}) = 64 \text{ cm}^2$

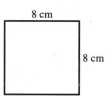

4. $S = 28 \text{ in.} \div 4 = 7 \text{ in.}$
$A = (7 \text{ in.}) \times (7 \text{ in.}) = 49 \text{ in.}^2$

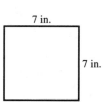

5. $S = (81 \text{ cm}^2)^{1/2} = 9 \text{ cm}$
$P = 4 \times (9 \text{ cm}) = 36 \text{ cm}$

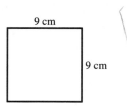

6. $S = 14 \text{ in.} \div 4 = 3.5 \text{ in.}$
$A = (3.5 \text{ in.}) \times (3.5 \text{ in.}) = 12.25 \text{ in.}^2$

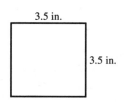

Exercise Set 17.2

1. $P = 2 \times (6 \text{ cm}) + 2 \times (14 \text{ cm}) = 40 \text{ cm}$
$A = (14 \text{ cm}) \times (6 \text{ cm}) = 84 \text{ cm}^2$

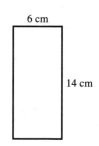

Exercise Set 17.2, *Cont.*

2. $P = 2 \times (5 \text{ in.}) + 2 \times (3 \text{ in.}) = 16 \text{ in.}$
    $A = (5 \text{ in.}) \times (3 \text{ in.}) = 15 \text{ in.}^2$

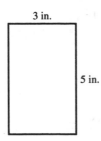

3. $S = 40 \text{ cm}^2 \div 5 \text{ cm} = 8 \text{ cm}$
    $P = 2 \times (5 \text{ cm}) + 2 \times (8 \text{ cm}) = 26 \text{ cm}$

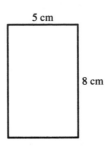

4. $P = 2 \times (5.3 \text{ cm}) + 2 \times (2.7 \text{ cm}) = 16 \text{ cm}$
    $A = (5.3 \text{ cm}) \times (2.7 \text{ cm}) = 14.31 \text{ cm}^2$

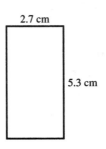

Exercise Set 17.3

1. $P = 9 \text{ cm} + 7 \text{ cm} + 4 \text{ cm} = 20 \text{ cm}$
    $A = (1/2) \times (9 \text{ cm}) \times (2.981 \text{ cm}) = 13.415 \text{ cm}^2$

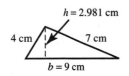

2. $P = 10 \text{ in.} + 10 \text{ in.} + 16 \text{ in.} = 36 \text{ in.}$
    $A = (1/2) \times (16 \text{ in.}) \times (6 \text{ in.}) = 48 \text{ in.}^2$

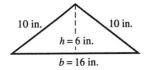

Exercise Set 17.3, *Cont.*

3. $P = 4.72 \text{ cm} + 8.61 \text{ cm} + 6.43 \text{ cm} = 19.76 \text{ cm}$
$A = (1/2) \times (8.61 \text{ cm}) \times (3.47 \text{ cm}) = 14.938 \text{ cm}^2$

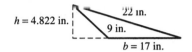

4. $P = 17 \text{ in.} + 9 \text{ in.} + 22 \text{ in.} = 48 \text{ in.}$
$A = (1/2) \times (17 \text{ in.}) \times (4.822 \text{ in.}) = 40.987 \text{ in.}^2$

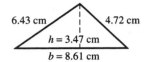

Exercise Set 17.4

1. $P = 2 \times (24 \text{ in.}) + 2 \times (18 \text{ in.}) = 84 \text{ in.}$
$A = (24 \text{ in.}) \times (15 \text{ in.}) = 360 \text{ in.}^2$

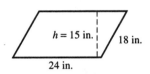

2. $P = 2 \times (4.6 \text{ cm}) + 2 \times (7.8 \text{ cm}) = 24.8 \text{ cm}$
$A = (7.8 \text{ cm}) \times (3.2 \text{ cm}) = 24.96 \text{ cm}^2$

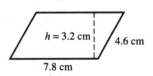

3. $P = 2 \times (3.4 \text{ ft}) + 2 \times (6.7 \text{ ft}) = 20.2 \text{ ft}$
$A = (6.7 \text{ ft}) \times (2.6 \text{ ft}) = 17.42 \text{ ft}^2$

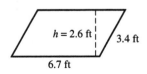

4. $P = 2 \times (5.7 \text{ mm}) + 2 \times (9.3 \text{ mm}) = 30 \text{ mm}$
$A = (9.3 \text{ mm}) \times (4.4 \text{ mm}) = 40.92 \text{ mm}^2$

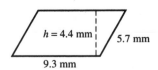

Exercise Set 17.5

1. $C = 2\pi(7 \text{ cm}) = 14\pi \text{ cm} = 43.982 \text{ cm}$
   $A = \pi(7 \text{ cm})^2 = 49\pi \text{ cm}^2 = 153.938 \text{ cm}^2$

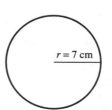

2. $C = 18\pi \text{ in.} = 56.549 \text{ in.}$
   $A = \pi(9 \text{ in.})^2 = 81\pi \text{ in.}^2 = 254.47 \text{ in.}^2$

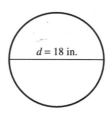

3. $C = 19.72\pi \text{ cm} = 61.952 \text{ cm}$
   $A = \pi(9.86 \text{ cm})^2 = 305.42 \text{ cm}^2$

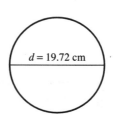

4. $d = C/\pi = 34.8 \text{ cm}/\pi = 11.077 \text{ cm}$
   $r = 11.077 \text{ cm}/2 = 5.5385 \text{ cm}$
   $A = \pi(5.5385 \text{ cm})^2 = 96.368 \text{ cm}^2$

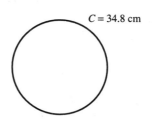

5. $d = C/\pi = 81.3 \text{ in.}/\pi = 25.879 \text{ in.}$
   $r = 25.879 \text{ in.}/2 = 12.939 \text{ in.}$
   $A = \pi(12.939 \text{ in.})^2 = 525.98 \text{ in.}^2$

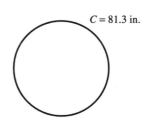

Exercise Set 17.6

1. $A = 6 \times (5 \text{ in.})^2 = 150 \text{ in.}^2$
   $V = (5 \text{ in.})^3 = 125 \text{ in.}^3$

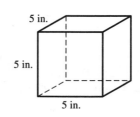

Exercise Set 17.6, *Cont.*

2. $A = 6 \times (7 \text{ cm})^2 = 294 \text{ cm}^2$
   $V = (7 \text{ cm})^3 = 343 \text{ cm}^3$

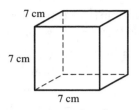

3. $S^2 = A/6 = 64 \text{ cm}^2/6 = 10.6\dots \text{ cm}^2$
   $S = (10.6\dots \text{ cm}^2)^{1/2} = 3.266 \text{ cm}$
   $V = (3.266 \text{ cm})^3 = 34.837 \text{ cm}^3$

4. $S = (V)^{1/3} = (216 \text{ cm}^3)^{1/3} = 6 \text{ cm}$
   $A = 6 \times (6 \text{ cm})^2 = 216 \text{ cm}^2$

5. $S^2 = A/6 = (625 \text{ in.}^2)/6 = 104.16\dots \text{ in.}^2$
   $S = (104.6\dots \text{ in.}^2)^{1/2} = 10.206 \text{ in.}$
   $V = (10.206 \text{ in.})^3 = 1063.1 \text{ in.}^3$

6. $S = (343 \text{ in.}^3)^{1/3} = 7 \text{ in.}$
   $6s^2 = 294 \text{ in.}^2$

7. $A = 6 \times (8 \text{ in.})^2 = 384 \text{ in.}^2$
   $V = (8 \text{ in.})^3 = 512 \text{ in.}^3$

Exercise Set 17.7

1. $A_1 = (8 \text{ in.}) \times (6 \text{ in.}) = 48 \text{ in.}^2$
   $A_2 = (6 \text{ in.}) \times (12 \text{ in.}) = 72 \text{ in.}^2$
   $A_3 = (12 \text{ in.}) \times (8 \text{ in.}) = 96 \text{ in.}^2$
   Total surface area $= 2A_1 + 2A_2 + 2A_3 = 432 \text{ in.}^2$
   $V = (8 \text{ in.}) \times (6 \text{ in.}) \times (12 \text{ in.}) = 576 \text{ in.}^3$

2. $A_1 = (9.3 \text{ cm}) \times (8.1 \text{ cm}) = 75.33 \text{ cm}^2$
   $A_2 = (8.1 \text{ cm}) \times (6.6 \text{ cm}) = 53.46 \text{ cm}^2$
   $A_3 = (6.6 \text{ cm}) \times (9.3 \text{ cm}) = 61.38 \text{ cm}^2$
   Total surface area $= 2A_1 + 2A_2 + 2A_3 = 380.34 \text{ cm}^2$
   $V = (9.3 \text{ cm}) \times (8.1 \text{ cm}) \times (6.6 \text{ cm}) = 497.2 \text{ cm}^3$

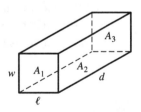

3. $A_1 = 49 \text{ cm}^2 = (7 \text{ cm}) \times (7 \text{ cm})$
   $A_1 = (7 \text{ cm}) \times (7 \text{ cm}) = 49 \text{ cm}^2$
   $A_2 = (7 \text{ cm}) \times (9 \text{ cm}) = 63 \text{ cm}^2$
   $A_3 = (9 \text{ cm}) \times (7 \text{ cm}) = 63 \text{ cm}^2$
   Total surface area $= 2A_1 + 2A_2 + 2A_3 = 350 \text{ cm}^2$
   $V = (7 \text{ cm}) \times (7 \text{ cm}) \times (9 \text{ cm}) = 441 \text{ cm}^3$

4. $A_1 = 63 \text{ cm}^2 = (7 \text{ cm}) \times (9 \text{ cm})$
   $A_1 = (7 \text{ cm}) \times (9 \text{ cm}) = 63 \text{ cm}^2$
   $A_2 = (7 \text{ cm}) \times (12 \text{ cm}) = 84 \text{ cm}^2$
   $A_3 = (12 \text{ cm}) \times (9 \text{ cm}) = 108 \text{ cm}^2$
   Total surface area $= 2A_1 + 2A_2 + 2A_3 = 510 \text{ cm}^2$
   $V = (7 \text{ cm}) \times (9 \text{ cm}) \times (12 \text{ cm}) = 756 \text{ cm}^3$

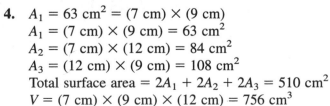

Exercise Set 17.8

1. Surface area $= 4\pi(5 \text{ cm})^2 = 100\pi \text{ cm}^2 = 314.16 \text{ cm}^2$
   $V = (4/3)\pi(5 \text{ cm})^3 = 16.666... \pi \text{ cm}^3 = 523.60 \text{ cm}^3$

2. Surface area $= 4\pi(4 \text{ in.})^2 = 64\pi \text{ in.}^2 = 201.06 \text{ in.}^2$
   $V = (4/3)\pi(4 \text{ in.})^3 = 85.333... \pi \text{ in.}^3 = 268.08 \text{ in.}^3$

3. Surface area $= 3\pi(8 \text{ cm})^2 = 192\pi \text{ cm}^2 = 603.2 \text{ cm}^2$
   $V = (2/3)\pi(8 \text{ cm})^3 = 314.33\pi \text{ cm}^3 = 1072.3 \text{ cm}^3$

4. Surface area $= 3\pi(14 \text{ in.})^2 = 588\pi \text{ in.}^2 = 1847.3 \text{ in.}^2$
   $V = (2/3)\pi(14 \text{ in.})^3 = 1829.3... \pi \text{ in.}^3 = 5747 \text{ in.}^3$

Exercise Set 17.9

1. Surface area $= 2\pi(5 \text{ cm})^2 + 2\pi(5 \text{ cm}) \times (9 \text{ cm})$
   $= 157.08 \text{ cm}^2 + 282.74 \text{ cm}^2 = 439.82 \text{ cm}^2$
   $V = \pi(5 \text{ cm})^2 \times (9 \text{ cm}) = 706.858 \text{ cm}^3$

2. Surface area $= 2\pi(4 \text{ in.})^2 + 2\pi(4 \text{ in.}) \times (8 \text{ in.})$
   $= 100.53 \text{ in.}^2 + 201.06 \text{ in.}^2 = 301.593 \text{ in.}^2$
   $V = \pi(4 \text{ in.})^2 \times (8 \text{ in.}) = 402.124 \text{ in.}^3$

3. Surface area $= 2\pi(12.5 \text{ cm})^2 + 2\pi(12.5 \text{ cm}) \times (17 \text{ cm})$
   $= 981.75 \text{ cm}^2 + 1335.18 \text{ cm}^2 = 2316.93 \text{ cm}^2$
   $V = \pi(12.5 \text{ cm})^2 \times (17 \text{ cm}) = 8344.86 \text{ cm}^3$

4. Surface area $= 2\pi(75.6 \text{ in.})^2 + 2\pi(75.6 \text{ in.}) \times (52.11 \text{ in.})$
   $= 35\ 910.7 \text{ in.}^2 + 24\ 752.7 \text{ in.}^2 = 60\ 663.4 \text{ in.}^2$
   $V = \pi(75.6 \text{ in.})^2 \times (52.11 \text{ in.}) = 935\ 652.4 \text{ in.}^3$

CHAPTER 18

Exercise Set 18.1

1. **a.** Mean average $= \$657\ 000/9$ houses
   $= \$73\ 000$ per house
   Median average $= \$55\ 000$ per house
   Mode average $= \$40\ 000$ per house

b.

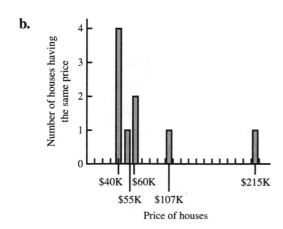

Exercise Set 18.1, *Cont.*

2. a.

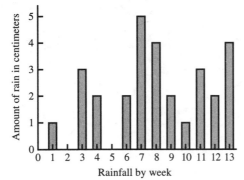

Rainfall by week

b. Mean average = 29 cm/13 weeks
= 2.231 cm/week
Median average = 2 cm/week of rainfall
Mode average = 2 cm/week of rainfall

c.

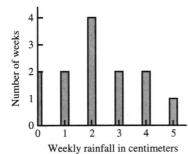

Weekly rainfall in centimeters

3. a. Mean average = 306 cm/17 trees
= 18 cm/tree
Median average = 17 cm/tree
Mode average = 15 cm/tree

b.

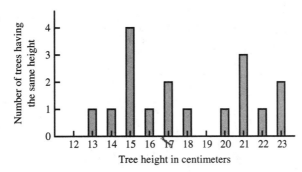

Tree height in centimeters

Exercise Set 18.1, *Cont.*

4. a. Mean average = 187 passengers/16 buses
= 11.688 passengers/bus
Median average = 14 passengers/bus
Mode average = 16 passengers/bus

b.

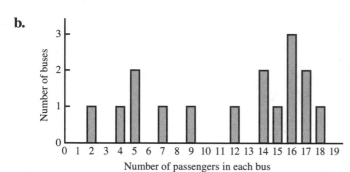

Exercise Set 18.2

1. a. 13.043% red buttons, 24.08% orange buttons, 17.726% yellow buttons, 21.40% green buttons, 8.361% blue buttons, 15.385% brown buttons
   b. Total = 99.995%, which is close to 100%
2. a. 11.64% red buttons, 18.10% orange buttons, 26.29% yellow buttons, 6.03% green buttons, 14.22% blue buttons, 23.71% brown buttons
   b. Total = 99.99%, which is close to 100%

Exercise Set 18.3

1. a.

Price of House	Number of Houses	Percent of Each Price
$ 40 000	4	44.444...%
$ 55 000	1	11.111...%
$ 60 000	2	22.222...%
$107 000	1	11.111...%
$215 000	1	11.111...%

Exercise Set 18.3, *Cont.*

b.

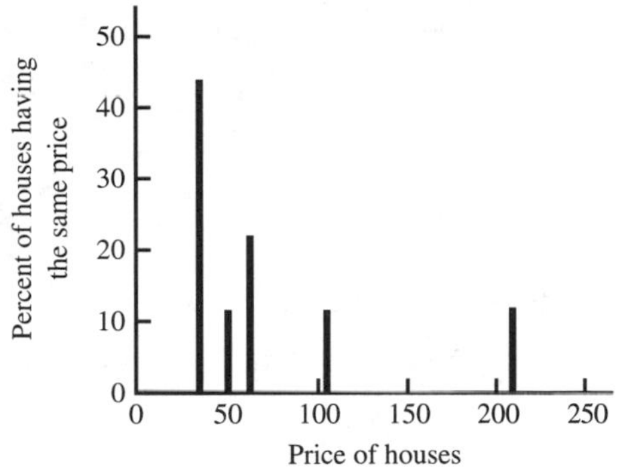

Percent of houses having the same price
50
40
30
20
10
0
0
50
100
150
200
250
Price of houses

2. a.

Tree Height in Centimeters	Number of Trees	Percent of Living Trees of Same Height
13	1	5.882%
14	1	5.882%
15	4	23.530%
16	1	5.882%
17	2	11.765%
18	1	5.882%
20	1	5.882%
21	3	17.647%
22	1	5.882%
23	2	11.765%

b. 85% of trees lived.
c. 15% of trees died.

3.

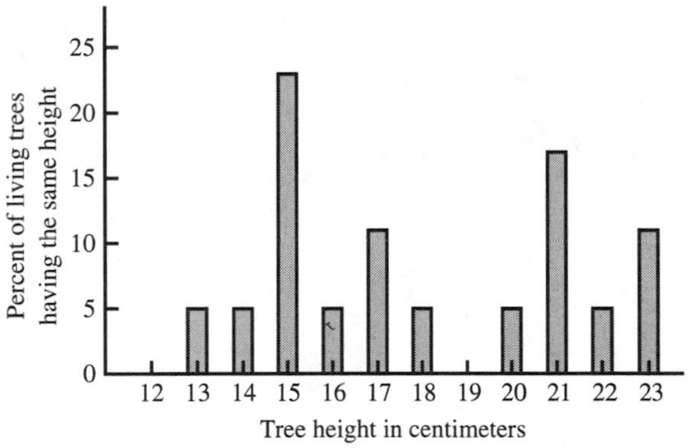

Percent of living trees having the same height
25
20
15
10
5
0
12 13 14 15 16 17 18 19 20 21 22 23
Tree height in centimeters

Exercise Set 18.3, *Cont.*

4.

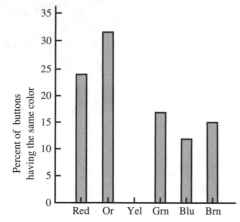

5. **a.** Mean average = 61 children/17 houses
$$= 3.588 \text{ children per house}$$
Median average = 3 children per house
Mode averages = 2 and 3 children per house

b.

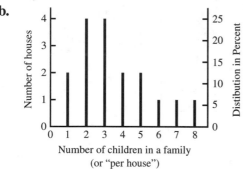

Exercise Set 18.4

1. **a.** Sum of areas (in dollars)
$$= 6 \times 20 + 2 \times 10 + 2 \times (-15) + 4 \times 10 + 2 \times 20 + 2 \times 30$$
$$= \$250$$

b. Area ÷ base = \$250 ÷ 20 weeks = \$12.50/week

c.

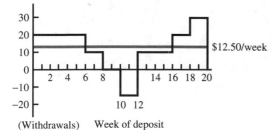

CHAPTER 19

Exercise Set 19.1

1. Slope = 4 ft/8 ft = 1/2 or 0.5
2. Slope = 352 cm/560 cm = $(2 \times 11)/(5 \times 7)$
   = 22/35 or 0.6286
3. Slope = 75 cm/4000 cm
   = 3/160 = 0.018 75
   Grade = slope $\times$ 100
   = 1.875%
4. Slope = -2 ft/80 ft
   = $-1/40$ = -0.025
   Grade = -0.025×100 = -2.5%

Exercise Set 19.2

1. $2 - (-4) = 2 + 4 = 6$
2. $3 - (-5) = 3 + 5 = 8$
3. $3 - (-5) = 3 + 5 = 8$
4. $2 - (-4) = 2 + 4 = 6$

Exercise Set 19.3

1. **a.**

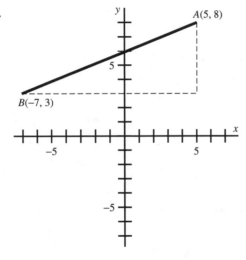

   b. Horizontal distance = $5 - (-7) = 5 + 7 = 12$
   c. Vertical distance = $8 - 3 = 5$
   d. Slope = 5/12 = 0.416...

Exercise Set 19.3, *Cont.*

2. a.

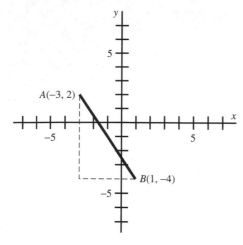

b. Horizontal distance = $1 - (-3) = 1 + 3 = 4$
c. Vertical distance = $(-4) - 2 = -4 - 2 = -6$
d. Slope = $-6/4 = -1.5$

3. a.

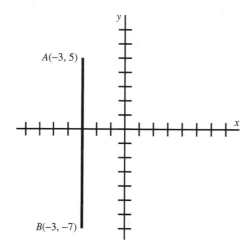

b. Horizontal distance = $-3 - (-3) = -3 + 3 = 0$
c. Vertical distance = $5 - (-7) = 5 + 7 = 12$
d. Slope = $12/0$ = undefined

Exercise Set 19.3, _Cont._

4. **a.**

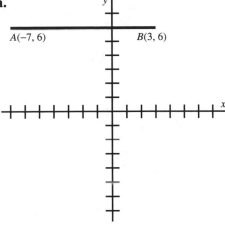

    b. Horizontal distance $= 3 - (-7) = 3 + 7 = 10$
    c. Vertical distance $= 6 - 6 = 0$
    d. Slope $= 0/10 = 0$

5. **a.**

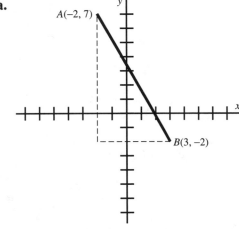

    b. Horizontal distance $= -2 - 3 = -5$
    c. Vertical distance $= 7 - (-2) = 7 + 2 = 9$
    d. Slope $= 9/(-5) = -1.8$

Exercise Set 19.3, *Cont.*

6. a.

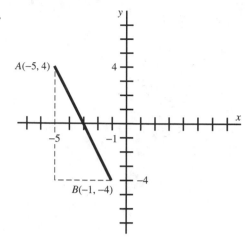

b. Horizontal distance $= -5 - (-1) = -5 + 1 = -4$
c. Vertical distance $= 4 - (-4) = 4 + 4 = 8$
d. Slope $= 8/(-4) = -2$

Exercise Set 19.4

1. a. Hypotenuse or $\Delta s = ((6 \text{ in.})^2 + (8 \text{ in.})^2)^{1/2} = 10 \text{ in.}$

b.

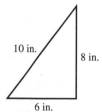

2. a. Hypotenuse or $\Delta s = ((13.6 \text{ cm})^2 + (9.4 \text{ cm})^2)^{1/2} = 16.532 \text{ cm}$

b.

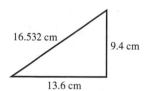

3. a. Hypotenuse or $\Delta s = ((5 \text{ cm})^2 + (12 \text{ cm})^2)^{1/2} = 13 \text{ cm}$

b.

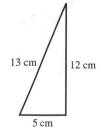

c. Slope $= 12 \text{ cm}/5 \text{ cm} = 2.4$

Exercise Set 19.4, *Cont.*

4. **a.** Hypotenuse or $\Delta s = ((32.4 \text{ in.})^2 + (21.8 \text{ in.})^2)^{1/2} = 39.05$ in.

  b.

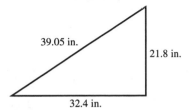

39.05 in.   21.8 in.

32.4 in.

  c. Slope = 21.8 in./32.4 in. = 0.6728

5. **a.**

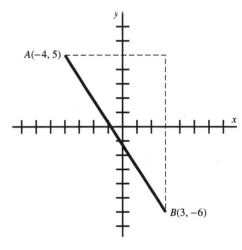

$A(-4, 5)$

$B(3, -6)$

  b. Horizontal distance = $-4 - 3 = -7$
  c. Vertical distance = $5 - (-6) = 5 + 6 = 11$
  d. Slope = $11/(-7) = -1.5714$
  e. $\Delta s = \sqrt{(-7)^2 + (11^2)} = \sqrt{170} = 13.038$

6. **a.**

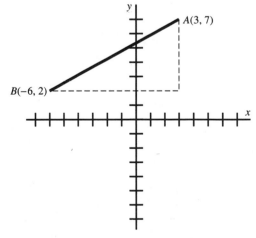

$A(3, 7)$

$B(-6, 2)$

  b. Horizontal distance = $3 - (-6) = 9$
  c. Vertical distance = $7 - 2 = 5$
  d. Slope = $5/9 = 0.5\ldots$
  e. $\Delta s = \sqrt{(9)^2 + (5)^2} = \sqrt{106} = 10.296$

Exercise Set 19.5

1. $\Delta s = \sqrt{(5)^2 + (2)^2 + (4)^2}$
    $= \sqrt{45} = 6.708$

2. $\Delta s = \sqrt{(8.2 \text{ cm})^2 + (-5.6 \text{ cm})^2 + (9.3 \text{ cm})^2}$
    $= \sqrt{185.09 \text{ cm}^2} = 13.605 \text{ cm}$

3. $\Delta s = \sqrt{(-7)^2 + (5)^2 + (-4)^2}$
    $= \sqrt{90} = 9.4868$

4. $\Delta s = \sqrt{(-9.6 \text{ cm})^2 + (-7.8 \text{ cm})^2 + (6.7 \text{ cm})^2}$
    $= \sqrt{197.89 \text{ cm}^2} = 14.067 \text{ cm}$

Exercise Set 19.6

1. **a.** $\Delta s = 40$ km east; $\Delta t = 4$ h
    $v = (40 \text{ km east})/(4 \text{ h}) = 10$ km/h east

    b.

Δt (in hours)	Δs (in km)	v (in km/h)
0	0	10
1	10	10
2	20	10
3	30	10
4	40	10

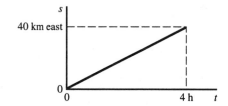

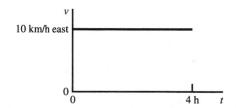

Exercise Set 19.6, *Cont.*

2. **a.** $\Delta t = 5$ h; $v = 5$ km/h north
    $\Delta s = v\,\Delta t = (5$ km/h north$)(5$ h$) = 25$ km north at end of trip

    b.

Δt (in hours)	Δs (in km)	v (in km/h)
0	0	˙5
1	5	5
2	10	5
3	15	5
4	20	5
5	25	5

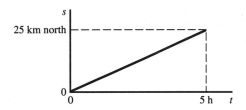

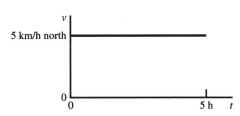

Exercise Set 19.7

1. **a.** Runner's distance from the reference point after 20 seconds:
$$s = vt + s_0 = [(10 \text{ m/s north}) \times (20 \text{ s}) + 40 \text{ m north}]$$
$$= [200 \text{ m north} + 40 \text{ m north}] = 240 \text{ m north}$$

b.

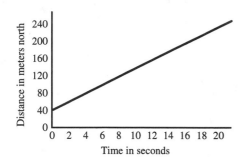

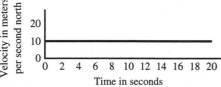

2. **a.** and **c.**

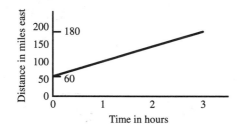

b. $v = \Delta s / \Delta t = (180 \text{ mi east} - 60 \text{ mi east})/(3 \text{ hr})$
$$= 40 \text{ mi/hr east}$$

Exercise Set 19.8

1. **a.** and **c.**

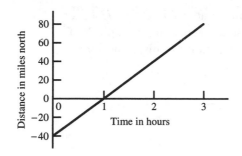

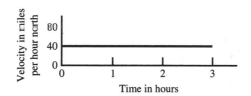

b. $v = \Delta s / \Delta t = [(80 \text{ mi} - (-40 \text{ mi}))/(3 \text{ hr})]$ north
$= (120 \text{ mi north})/(3 \text{ hr}) = 40 \text{ mi/hr north}$

d. The car will pass the reference point in exactly one hour.
Verify your answer by substituting this value into the distance
formula:

$$s = vt + s_0$$
$$s = (40 \text{ mi/hr north})(1 \text{ hr}) + (-40 \text{ mi})$$
$$= 40 \text{ mi} - 40 \text{ mi} = 0 \text{ mi}$$

2. **a.** and **c.**

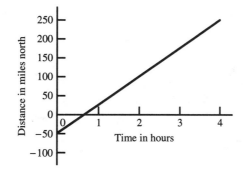

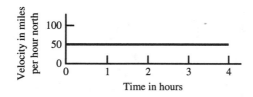

b. $v = \Delta s / \Delta t = [(250 \text{ mi} - (-50 \text{ mi})/(4 \text{ hr})]$ north
$= (300 \text{ mi north})/(4 \text{ hr}) = 75 \text{ mi/hr north}$

Exercise Set 19.8, *Cont.*

d. The car will pass the reference point in approximately 2/3 hr, or 40 minutes—see the preceding distance-versus-time graph. You may verify your approximation by substituting this value into the distance formula:

$$s = vt + s_0$$
$$= (75 \text{ mi/hr north})(2/3 \text{ hr}) + (-50 \text{ mi})$$
$$= 50 \text{ mi} - 50 \text{ mi}$$
$$= 0 \text{ miles}$$

Exercise Set 19.9

1. **a.** and **b.**

t	s	v	a
0	0	0	4
1	2	4	4
2	8	8	4
3	18	12	4
4	32	16	4
5	50	20	4

c.

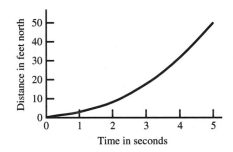

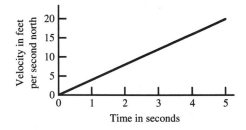

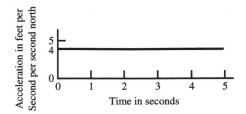

Exercise Set 19.9, *Cont.*

2. **a.** and **b.**

t	s	v	a
0	0	0	6
1	3	6	6
2	12	12	6
3	27	18	6
4	48	24	6
5	75	30	6

c.

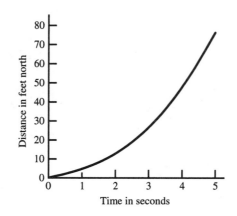

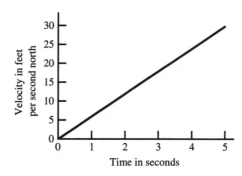

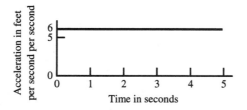

CHAPTER 20

Exercise Set 20.1

1. **a.** 2, 4, 6, 8, 10, 12, 14, 16
    b. 2 + 4 + 6 + 8 + 10 + 12 + 14 + 16
    c. 72

Exercise Set 20.1, *Cont.*

2. **a.** $1, -3, 5, -7, 9, -11, 13$
   b. $1 - 3 + 5 - 7 + 9 - 11 + 13$
   c. 7

3. **a.** $3, 6, 12, 24, 48, 96$; thus 96
   b. and **c.** $3 + 6 + 12 + 24 + 48 + 96$
   d. 189

4. **a.** 54 in.
   b. $16, -8, +24, -12, +36, -18, +54$
   c. 54

5. **a.** $160
   b. $50, $20, $40, $50, $30, $60, $10
   c. $260

Exercise Set 20.2

1. $1/1 + 1/2 + 1/3 + 1/4 = 2.083\ldots$
2. $3 + 5 + 7 + 9 + 11 = 35$
3. $1 - 2 + 4 - 8 + 16 = 11$
4. $9 + 27 + 81 + 243 = 360$
5. $2/3 + 3/5 + 4/7 = 1.8381$
6. $1/4 + 1/2 + 1 + 2 + 4 + 8 = 15.75$

Exercise Set 20.3

1. **a.** $a_1 = 2, d = 3, a_2 = 5, a_3 = 8, a_4 = 11, a_5 = 14, a_6 = 17$
   b. $S_6 = (6/2)(2 + 17) = 57$

   c.

2. **a.** $a_1 = 3, d = 1.5, a_2 = 4.5, a_3 = 6, a_4 = 7.5, a_5 = 9$
   b. $S_5 = (95/2)(3 + 9) = 30$

   c.

3. **a.** $a_1 = 11, d = -4, a_2 = 7, a_3 = 3, a_4 = -1, a_5 = -5$
   b. $S_5 = (5/2)[11 + (-5)] = 15$

   c.

Exercise Set 20.3, *Cont.*

4. **a.** $a_1 = -5$, $a_2 = -1$, $d = 4$, $a_3 = 3$, $a_4 = 7$, $a_5 = 11$

   b. $S_5 = (5/2)(-5 + 11) = 15$

   c.

Exercise Set 20.4

1. **a.** $a_1 = 5$, $r = 2$, $a_2 = 5 \times 2^{2-1} = 10$, $a_3 = 5 \times 2^{3-1} = 20$

   b. $S_3 = (5 - 2 \times 20)/(1 - 2) = 35$

2. **a.** $a_1 = 5$, $a_2 = 20$, $r = 20/5 = 4$, $a_3 = 5 \times 4^{3-1} = 80$,
   $a_4 = 5 \times 4^{4-1} = 320$

   b. $S_4 = (5 - 4 \times 320)/(1 - 4) = 425$

Exercise Set 20.5

1. For the left decade (0.1 to 1), one spacing difference is $0.5 - 0.1 = 0.4$, and the ratio is $0.5/0.1 = 5$.

   For the center decade (1 to 10), one spacing difference is $5 - 1 = 4$, and the ratio is $5/1 = 5$.

   For the right decade (10 to 100), one spacing difference is $50 - 10 = 40$, and the ratio is $50/10 = 5$.

   Note that the differences are not the same; however, the ratios are identical (5).

2. **a.** For the left decade (0.1 to 1), one space difference is 0.1, 0.3, and 0.5, and the ratios are 2/1, 2.5/1, and 2/1.

   For the center decade (1 to 10), one space difference is 1, 3, and 5, and the ratios are 2/1, 2.5/1, and 2/1.

   For the right decade (10 to 100), one space difference is 10, 30, and 50, and the ratios are 2/1, 2.5/1, and 2/1.

   b. Note that the differences are not the same; however, the three ratios within each decade are identical (2/1, 2.5/1, and 5/1).

Exercise Set 20.6

1. **a.** $a_1 = 4$, $r = 5$, $a_2 = 4 \times 5^{2-1} = 20$, $a_3 = 4 \times 5^{3-1} = 100$,
   $a_4 = 4 \times 5^{4-1} = 500$, $a_5 = 4 \times 5^{5-1} = 2500$

   b. $S_5 = (4 - 5 \times 2500)/(1 - 5) = 3124$

   c.

Exercise Set 20.6, *Cont.*

2. **a.** $a_1 = 3$, $a_2 = 30$, $r = 30/3 = 10$, $a_3 = 3 \times 10^{3-1} = 300$,
$a_4 = 3 \times 10^{4-1} = 3\ 000$, $a_5 = 3 \times 10^{5-1} = 30\ 000$,
$a_6 = 3 \times 10^{6-1} = 300\ 000$

b. $S_6 = (3 - 10 \times 300\ 000)/(1 - 10) = 333\ 333$

c.

3. **a.** $a_1 = 2$, $a_2 = 20$, $r = 20/2 = 10$, $a_3 = 2 \times 10^{3-1} = 200$,
$a_4 = 2 \times 10^{4-1} = 2000$, $a_5 = 2 \times 10^{5-1} = 20\ 000$

b. $S_5 = (2 - 10 \times 20\ 000)/(1 - 10) = 22\ 222$

c.

4.

Musical Note	Frequency
Two octaves below	130.75 Hz
One octave below	261.5 Hz
Middle C	523 Hz
One octave above	1046 Hz
Two octaves above	2092 Hz
Three octaves above	4184 Hz
Four octaves above	8368 Hz

Exercise Set 20.7

1.

Date (Year)	Deposit at Beginning of Year	Interest Earned at End of Year
0	$10.00	$0.60
1	$10.60	$0.64
2	$11.24	$0.67
3	$11.91	$0.71
4	$12.62	$0.76
5	$13.38	$0.80
6	$14.19	$0.85
7	$15.04	$0.90
8	$15.94	$0.96
9	$16.89	$1.01
10	$17.91	$1.07
11	$18.98	$1.14
12	$20.12	

Note: Twelve years is required.

MATH AT WORK

Exercise Set 20.7, Cont.

2.

Number of Years Required	Millions of Gallons Used per Year	End-of-Year Increase in Gallons
0	100	14
1	114	15.96
2	129.96	18.19
3	148.15	20.74
4	168.90	23.65
5	192.54	26.96
6	219.50	30.73
7	250.23	35.03
8	285.26	39.94
9	325.19	45.53
10	370.72	

Final two years: 695.91 millions of gallons is estimated to be used. The reservoir should contain approximately 700M gallons.

3.

Date (Year)	Amount of Oil Extracted (M Barrels)	Total Oil Extracted (M Barrels)
1	6	6
2	5.58	11.58
3	5.19	16.77
4	4.83	21.60
5	4.49	26.09
6	4.18	30.27
7	3.89	34.16
8	3.62	37.78

Note: The oilfield will be able to supply oil for eight years, assuming that the original estimate is correct.

4.

Year	Price Today	Price Increase
0	$150 000.00	$11 700.00
1	$161 700.00	$12 612.60
2	$174 312.60	$13 596.38
3	$187 908.98	$14 656.90
4	$202 565.88	$15 800.14
5	$218 366.02	$17 032.55
6	$235 398.57	

Note: The value of the house after six years, based upon an estimated price-increase rate of 7.8%, is approximately $235 400.

CHAPTER 21

Exercise Set 21.1

1. **a.** $\theta = \Delta s / \Delta r = (1.25 \text{ in.})/(2 \text{ in.}) = 0.625$ radian

   b.

   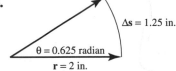

   $\Delta s = 1.25$ in.

   $\theta = 0.625$ radian

   $r = 2$ in.

2. **a.** $\theta = \Delta s / \Delta r = (2 \text{ cm})/(5 \text{ mm}) = 20 \text{ mm}/5 \text{ mm} = 4$ radians

   b.

   $\Delta s = 2$ cm

   $\theta = 4$ radians

   $r = 5$ mm

3. **a.** $\theta = \Delta s / \Delta r = (-10 \text{ in.})/(5 \text{ in.}) = -2$ radians

   b.

   $r = 5$ in.

   $\theta = -2$ radians

   $\Delta s = -10$ in.

4. $s = 8 \text{ in.} \times 2 \text{ radians} = 16 \text{ in.}$
5. $s = 12 \text{ cm} \times 0.5 \text{ radian} = 6 \text{ cm}$

Exercise Set 21.2

1. $2 \times 2 \times 2 \times 3 \times 3 \times 5$
2. **a.** $45° \times (2\pi \text{ radians}/360°) = 90\pi/360 \text{ radians} = \pi/4$ radians

   b.

   $45°$

   c. This angle is located in the first quadrant (I).

Exercise Set 21.2, *Cont.*

3. **a.** $(\pi/6)$ radians $\times$ $(360°/2\pi$ radians$) = 30°$

     b.

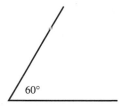

     c. This angle is located in the first quadrant (I).

4. **a.** $(1/6)$ revolution $\times$ $(360°/1$ revolution$) = 60°$

     b.

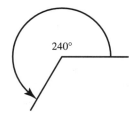

     c. This angle is located in the first quadrant (I).

5. **a.** $(2/3)$ revolution $\times$ $(360°/1$ revolution$) = 240°$

     b.

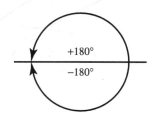

     c. This angle is located in the third quadrant (III).

6. $(2/3)$ revolution $\times$ $(2\pi$ radians$/1$ revolution$) = 4\pi/3$ radians

7. **a.** and **b.**

     $+180°$
     $-180°$

8. $180° \times (2\pi$ radians$/360°) = \pi$ radians

Exercise Set 21.3

1. a. $820° - 360° - 360° = 100°$

b.

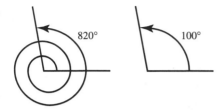

c. This angle is located in the second quadrant (II).

2. a. $-680° + 360° + 360° = 40°$

b.

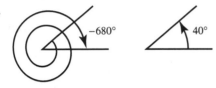

c. This angle is located in the first quadrant (I).

3. a. $1450° - 360° - 360° - 360° - 360° = 10°$

b.

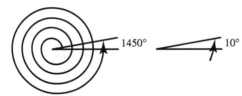

c. This angle is located in the first quadrant (I).

4. a. $-800° + 360° + 360° + 360° = 280°$

b.

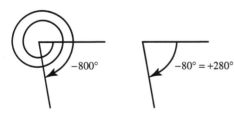

c. This angle is located in the fourth quadrant (IV).

Exercise Set 21.4

1. Opposite = (adjacent) $\times$ (tan θ) = (410 ft) $\times$ (tan 59.3°)

= (410 ft) $\times$ (1.6841919) = 690.5 ft

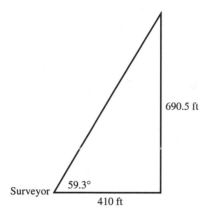

2. tan θ = 330 m/78 m; θ = invtan 330/78 = 76.70°

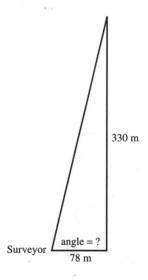

3. Vertical: Opposite = (hypotenuse) $\times$ (sin θ) = (50 m) $\times$ (sin 42°)

= (50 m) $\times$ (0.6691306) = 33.46 m

Horizontal: Adjacent = (hypotenuse) $\times$ (cos θ)

= (50 m) $\times$ (cos 42°)

= (50 m) $\times$ (0.7431448) = 37.16 m

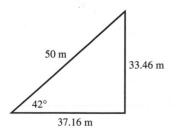

Exercise Set 21.4, *Cont.*

4. Hypotenuse = (opposite) ÷ (sin θ) = (2306 ft) ÷ (sin 36.52°)
= (2306 ft) ÷ (0.5951034) = 3875 ft

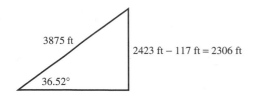

CHAPTER 22

Exercise Set 22.1

1. *Known:*
   2 offices
   6 persons in smaller office (*A*)
   12 persons in larger office (*B*)
   ages of each person
Unknown:
   average age in office *A*
   average age in office *B*
   average age in the building
Math:
   Add each set of ages; divide by the number of persons in that set.
Solution:
   a. Office *A*: (156 years ÷ 6 persons) = 26 years average per person
   b. Office *B*: (420 years ÷ 12 persons) = 35 years average per person
   c. Building: (576 years ÷ 18 persons) = 32 years average per person

2. *Known:*
   2 classrooms (*A* and *B*)
   15 students in classroom *A*; average age = 8.2 years
   21 students in classroom *B*; average age = 13 years
Unknown:
   average age of students in school, two classrooms
Math:
   Average = total of all ages (*A* and *B*) ÷ total number of students
Solution:
   Classroom *A* total of all ages: (8.2 yr/student × 15 students) = 123 years
   Classroom *B*: (13 yr/student × 21 students) = 273 years
   School: (123 + 273) yr ÷ (15 + 21) students = 11 years average per student

3. *Known:*
   average age and population for the inner city
   average age and population for each of the six suburbs
Unknown:
   average age for all six suburbs
   average age for metro Blatt
Math:
   Average = total of all ages for each section ÷ populations

Exercise Set 22.1, *Cont.*

Solution:

a. Suburbs: The total of all suburban ages (population × average ages) is

$$(250\ 000 \times 46.1) + (352\ 000 \times 32.4) + (145\ 000 \times 57.3)$$
$$+ (421\ 000 \times 45.2) + (85\ 000 \times 39.6) + (129\ 000 \times 51.7)$$
$$= 60\ 302\ 800$$

The total suburb population is

$$(250\ 000 + 352\ 000 + 145\ 000 + 421\ 000 + 85\ 000 + 129\ 000)$$
$$- 1\ 382\ 000$$

The average age for all six suburbs is

$$60\ 302\ 800\ \text{yr} \div 1\ 382\ 000\ \text{persons}$$
$$= 43.63\ \text{years (person's average age in the suburbs)}$$

b. Suburbs and city: The total of all ages is

$$(\text{Population} \times \text{average ages})$$
$$= (1\ 382\ 000 \times 43.63) + (1\ 600\ 000 \times 37.6)$$
$$= 120\ 462\ 800$$

The total suburb and city population is

$$1\ 382\ 000 + 1\ 600\ 000 = 2\ 982\ 000$$

The average age for metro Blatt is

$$120\ 462\ 800\ \text{yr} \div 2\ 982\ 000\ \text{persons}$$
$$= 40.397\ \text{years (average age per person)}$$

Exercise Set 22.2

1. *Known:*

Size of table = 13 ft × 7 ft
Tablecloth is to be 1 ft longer on each side.
Cloth costs $3.50 per square yard.
Widths of cloth available are 35″, 44″, 58″, and 60″.
Cloth is sold by the "running yard," which is one yard in length and whatever width is selected.

Unknown:

width and length of cloth
sq ft of cloth
sq ft of cloth to be purchased
total cost of the cloth

Math:

formula for the area of a rectangle

Exercise Set 22.2, *Cont.*

The sketch is as follows:

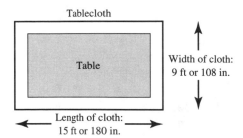

Solution:

a. Therefore, the size of the tablecloth in sq ft is

$$(15 \text{ ft}) \times (9 \text{ ft}) = 135 \text{ sq ft}$$

Note that 180 in. ÷ 3 is 60 in.; choose the 60-in.-width cloth.

b. Area in sq feet of cloth purchased: Purchase three 60-in. by 9-ft pieces, and sew the three pieces together.

$$3 \times ((60 \text{ in.} \times 1 \text{ foot}/12 \text{ in.}) \times 9 \text{ feet})$$
$$= 3 \times (5 \text{ feet} \times 9 \text{ feet}) = 3 \times 45 \text{ sq feet} = 135 \text{ sq feet}$$

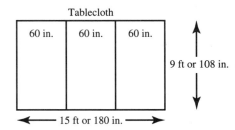

c. Cost per square yard of cloth purchased: Convert sq ft to sq yd; then calculate the cost.

$$135 \text{ sq ft} \times (1 \text{ yd}/3 \text{ ft}) \times (1 \text{ yd}/3 \text{ ft})$$
$$= 135 \text{ sq ft} \times 1 \text{ sq yd}/9 \text{ sq ft} = 15 \text{ sq yd}$$
$$15 \text{ sq yd} \times \$3.50/\text{sq yd} = \$52.50$$

You might take another approach.

Exercise Set 22.2, *Cont.*

Solution:

a. Note that 2×58 in. $= 116$ in. is slightly wider than the 108 in. required, as shown in the following figure.

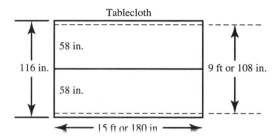

Tablecloth

Choose the 58-in.-wide cloth.

b. Area in sq feet of cloth purchased: Purchase two 58-in. by 15-ft pieces, trim 4 in. from each, and sew the two pieces together.

$$2 \times ((58 \text{ in.} \times 1 \text{ ft}/12 \text{ in.}) \times 15 \text{ ft})$$
$$= 2 \times (4.83 \text{ ft} \times 15 \text{ ft}) = 144.9 \text{ sq ft}$$

c. Cost per square yard of cloth purchased: Convert sq ft to sq yd; then calculate the cost.

$$144.9 \text{ sq ft} \times 1 \text{ yd}/3 \text{ ft} \times 1 \text{ yd}/3 \text{ ft}$$
$$= 144.9 \text{ sq ft} \times 1 \text{ sq yd}/9 \text{ sq ft} = 16.1 \text{ sq yd}$$

$$16.1 \text{ sq yd} \times \$3.50/\text{sq yd}$$
$$= \$56.35$$

2. *Known:*

A display consists of a 2-ft-square board and two 14-in.-diameter circles to be mounted on a square board, one on each surface.

Two coats of paint are to be applied to the display.

The first coat of paint is applied to both surfaces of the square board and to one surface of each of the circles. Both circles are then mounted on the square, and a second coat of paint is applied only to the visible portion of the square board.

The paint is sold by the pint can; it will cover 8 sq ft for the first coat and 12 sq ft for the second coat.

The paint costs \$8.50 per pint if only one can is purchased; it costs \$15.50 per quart if two or more cans are purchased at the same time.

Unknown:

the surface areas of the square board and the circles

the surface area to be painted for the first coat and the second coat

the number of cans of paint needed for the first coat, for the second coat (and, therefore, both coats together)

the total cost of the paint

Math:

formulas for the area of a square and a circle

use of the unity method to convert units

Exercise Set 22.2, *Cont.*

The sketch is as follows:

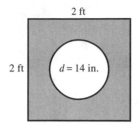

Solution:

Total surface area to be painted:

First coat:

Both surfaces of the square and one surface of each circle
$= 2 \times (2 \text{ ft})^2 + 2 \times (\pi \times (7 \text{ in.})^2)$
$= 2 \times 4 \text{ sq ft} + 2 \times (\pi \times (0.5833 \text{ ft})^2)$
$= 2 \times 4 \text{ sq ft} + 2 \times 1.069 \text{ sq ft}$
$= 8 \text{ sq ft} + 2.138 \text{ sq ft} = 10.138 \text{ sq ft}$

Second coat:

The difference between both square's surface areas and both circle's surface areas
$= 2 \times (4 \text{ sq ft} - 1.069 \text{ sq ft})$
$= 2 \times 2.931 \text{ sq ft} = 5.862 \text{ sq ft}$

a. Number of cans of paint needed for first coat:

$10.138 \text{ sq ft}/(8 \text{ sq ft/pint can}) = 1.27 \text{ pint cans}$

Number of cans of paint needed for second coat:

$5.862 \text{ sq ft}/(12 \text{ sq ft/pint can}) = 0.49 \text{ pint can}$

Therefore, the total amount of paint needed is

$(1.27 + 0.49) \text{ pint can} = 1.76 \text{ cans}$ or, at most, 2 pint cans

b. The total price of the paint is

$2 \text{ pints} \times \$15.50/\text{quart}$
$= 1 \text{ quart} \times \$15.50/\text{quart} = \$15.50$

3. *Known:*
   One table (*A*) has a radius of one yard and a circular hole cut in it whose diameter is 14 in.
   Another table (*B*) has a radius of one yard and a square hole cut in it whose side is 15 cm.
   Unknown:
   the areas of tabletops *A* and *B* remaining to be painted
   Math:
   formulas for the areas of a square and a circle
   use of the unity method to convert units

Exercise Set 22.2, *Cont.*

The sketch is as follows:

Tabletop *A*:

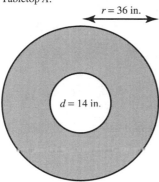

r = 36 in.

d = 14 in.

Tabletop *B*:

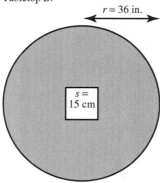

r = 36 in.

s = 15 cm

Solution:

a. The area of tabletop *A* to be painted is

Area of *A* − area of hole
= $\pi(36 \text{ in.})^2 - \pi(7 \text{ in.})^2$
= 4071.5 sq in. − 153.94 sq in. = 3917.56 sq in.

b. The area of tabletop *B* to be painted is

Area of *B* − area of hole
= $\pi(36 \text{ in.})^2 - [15 \text{ cm} \times (1 \text{ in.}/2.54 \text{ cm})]^2$
= 4071.5 sq in. − 34.88 sq in. = 4036.62 sq in.

Exercise Set 22.3

1. *Known:*
dimensions of the barn: 40 ft × 80 ft
starting point, ending point, and radius (10 ft) of the goat's circular walk
Unknown:
distance the goat has walked
square feet of grass available for the goat to eat
Math:
formulas for the circumference and area of a circle

Exercise Set 22.3, *Cont.*

The sketch is as follows:

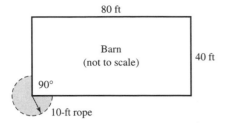

Solution:

a. The goat has walked part of the circumference of a circle that has a radius of 10 ft. How much of the circumference is this? The goat has walked 270 degrees. Therefore, the goat has walked 3/4 of the perimeter of the circle.

$$(3/4) \times 2 \, \pi \times (10 \text{ ft})$$
$$= 15\pi \text{ ft} = 47.12 \text{ ft (a distance)}$$

b. We also know that the goat has 3/4 of the area of a circle (which has a radius of 10 ft) of grass to eat. This area is

$$(3/4) \times \pi \times (10 \text{ ft})^2$$
$$= (3/4) \times \pi(100 \text{ sq ft})$$
$$= 75\pi \text{ sq ft} = 235.6 \text{ sq ft}$$

2. *Known:*

The speed of the U.S. runner is 1200 ft per 0.92 minute; the speed of the Romanian runner is 400 m per 62 seconds.

Unknown:

speed of both the U.S. and Romanian runners in miles per hour and in meters per second

Math:

use of the unity method to convert units

Solution:

a. The speed of the U.S. runner in mi/hr is

$$(1200 \text{ ft}/0.92 \text{ min}) \times (1 \text{ mi}/5280 \text{ ft}) \times (60 \text{ min}/1 \text{ hr})$$
$$= (1200 \times 60) \div (0.92 \times 5280) \text{ mi/hr}$$
$$= 14.822 \text{ mi/hr}$$

The speed of the Romanian runner in mi/hr is

$$(400 \text{ m}/62 \text{ s}) \times (100 \text{ cm}/1 \text{ m}) \times (1 \text{ in.}/2.54 \text{ cm}) \times (1 \text{ ft}/12 \text{ in.})$$
$$\times (1 \text{ mi}/5280 \text{ ft}) \times (60 \text{ sec}/1 \text{ min}) \times (60 \text{ min}/1 \text{ hr})$$
$$= (400 \times 100 \times 60 \times 60) \div (62 \times 2.54 \times 12 \times 5280) \text{ mi/hr}$$
$$= 14.432 \text{ mi/hr}$$

Exercise Set 22.3, *Cont.*

b. The speed of the U.S. runner in m/s is

$$(1200 \text{ ft/0.92 min}) \times (12 \text{ in./1 ft}) \times (2.54 \text{ cm/1 in.})$$
$$\times (1 \text{ m/100 cm}) \times (1 \text{ min/60 sec})$$
$$= (1200 \times 12 \times 2.54) \div (0.92 \times 100 \times 60) \text{ m/s}$$
$$= 6.626 \text{ m/s}$$

The speed of the Romanian runner in m/s is

$$(400 \text{ m/62 s})$$
$$= 6.452 \text{ m/s}$$

3. *Known:*

distances traveled in either miles or kilometers from Oshkosh (in the U.S.) to Ipsum (in Canada)

gas purchased (in either gallons or liters) between Oshkosh and Ipsum

Unknown:

total distance traveled between Oshkosh and Ipsum in miles and in kilometers

gas consumption between Oshkosh and Ipsum in miles/gal and in km/liter

Math:

use of the unity method to convert units

Solution:

a. Total distance traveled in miles: Combine as km.

$$(55 + 125 + 40 + 250 + 95 + 50 + 10) \text{ km} = 625 \text{ km}$$

Then convert km to miles.

$$625 \text{ km} \times (1000 \text{ m/1 km}) \times (100 \text{ cm/1 m}) \times (1 \text{ in./2.54 cm})$$
$$\times (1 \text{ ft/12 in.}) \times (1 \text{ mi/5280 ft})$$
$$= (625 \times 1000 \times 100) \div (2.54 \times 12 \times 5280) \text{ mi} = 388.36 \text{ mi}$$

Total miles traveled
$$= (150 + 60 + 25 + 40 + 388.6) \text{ mi} = 663.6 \text{ mi}$$

b. Total distance traveled in kilometers: Combine miles.

$$(150 + 60 + 25 + 40) \text{ mi} = 275 \text{ mi}$$

Convert mi to kilometers.

$$275 \text{ mi} \times (5280 \text{ ft/1 mi}) \times (12 \text{ in./1 ft}) \times (2.54 \text{ cm/1 in.})$$
$$\times (1 \text{ m/100 cm}) \times (1 \text{ km/1000 m})$$
$$= (275 \times 5280 \times 12 \times 2.54) \div (100 \times 1000) \text{ km} = 442.57 \text{ km}$$

Total kilometers traveled
$$= (442.57 + 625) \text{ km} = 1067.57 \text{ km}$$

Exercise Set 22.3, *Cont.*

c. Total gas consumption in miles/gallon: Combine liters and convert to gallons.

$$(36.2 + 29.7) \text{ liters} \times (0.264 \text{ gal/1 liter})$$
$$= (65.9 \times 0.264) \text{ gal} = 17.398 \text{ gal}$$

Total amount of gas consumed (in gal)
$$= (7.8 + 6.1 + 17.398) \text{ gal} = 31.298 \text{ gal}$$

Gas consumption in miles/gallon
= (total distance traveled in miles) ÷ (total gas consumed in gallons)
$$= (663.6 \text{ mi}) \div (31.298 \text{ gal}) = 21.20 \text{ mi/gal}$$

d. Total gas consumption in km/liter: Combine gallons and convert to liters.

$$(7.8 + 6.1) \text{ gal} \times (1 \text{ liter/0.264 gal}) = (13.9 \div 0.264) \text{ liters}$$
$$= 52.65 \text{ liters}$$

Total amount of gas consumed (in liters)
$$= (52.65 + 36.2 + 29.7) \text{ liters} = 118.55 \text{ liters}$$

Gas consumption in km/liter
= (total distance traveled in km ÷ total gas consumed in liters)
$$= (1067.57 \text{ km}) \div (118.55 \text{ liters}) = 9.005 \text{ km/liter}$$

Exercise Set 22.4

1. *Known:*
The sequence used is the odd integers, starting with 1.
Unknown:
Determine the first seven odd integers, display these on a real number line, and determine the series of this seven-number sequence.
Determine the eighty-second odd integer.
Determine the series of the odd integers, from the first (1) to the eighty-seventh.
Math:
The first seven odd integers are easily determined. Note that this is an arithmetic progression. Therefore, apply the formulas for determining any term in an arithmetic progression, as well as the value of any series of an arithmetic progression.
Solution:
a. The first seven odd integers are

$$1, \quad 3, \quad 5, \quad 7, \quad 9, \quad 11, \quad 13$$

(They may also be calculated by using the formulas.)

b. The display is as follows:

Exercise Set 22.4, *Cont.*

    c. The sum of these seven terms is

$$\Sigma = 1 + 3 + 5 + 7 + 9 + 11 + 13 = 49$$

The formula for the series of an arithmetic progression may be used.

$$S_7 = (7/2)(1 + 13) = (7/2)(14) = 49$$

    d. The 82nd odd integer, where $a_1 = 1$ and $d = 2$, is

$$a_{82} = 1 + (82 - 1) \times 2 = 1 + 81 \times 2 = 163$$

    e. The series from the first term to the 87th term requires the value of the 87th term (a_{87}), where $a_1 = 1$ and $d = 2$. Then

$$a_{87} = 1 + (87 - 1) \times 2 = 1 + 86 \times 2 = 173$$

Therefore, the value of the series of this 87th-term arithmetic progression is

$$S_{87} = (87/2)(1 + 173) = (87/2)(174) = 7569$$

2.   *Known:*

    Dan is 46 yr old.

    Dan is 10 yr older than John.

    Nancy is one-half the age of John.

    Dan earns twice as much money as John does.

    John earns one-third more money than Nancy does.

    Nancy earns $12 000 each year.

    John's house has a value of $90 000.

    John's house is worth 50% more than Dan's house.

    John's sailboat is worth $5200.

    Dan's sailboat is worth three times as much as Nancy's car.

    $5200 is worth 30% more than Nancy's car.

   Unknown:

    John's age and Nancy's age

    Dan's salary and John's salary

    the value of Dan's house and sailboat

    the value of Nancy's car

    the greater value, in percent, of Dan's boat compared to John's boat

   Math:

    addition and subtraction of integers

    multiplication and division of fractions and decimals

    percent

   Solution:

    John's age: Dan is 10 years older than John. Therefore, John is (46 − 10) years = 36 years old.

    Nancy's age: Nancy is one-half the age of John. Therefore, Nancy is 36/2 years = 18 years old.

John's salary (note that this salary must be determined before Dan's because Dan's salary has been compared to John's): John earns 1/3 more money than Nancy does ($12 000). Therefore, John earns Nancy's salary plus (1/3) × (Nancy's salary):

$$= \$12\ 000 + (1/3) \times (\$12\ 000)$$
$$= 1 \times \$12\ 000 + (1/3) \times (\$12\ 000)$$
$$= (4/3) \times (\$12\ 000)$$
$$= \$16\ 000$$

The problem could also have been solved as follows:

$$\$12\ 000 + (1/3) \times (\$12\ 000)$$
$$= \$12\ 000 + \$4000$$
$$= \$16\ 000$$

Dan's salary: Dan earns twice as much money as John earns. Therefore, Dan earns 2 × $16 000 = $32 000.

Dan's house: John's house ($90 000) is worth 50% more than Dan's house.

Value of John's house
$$= (\text{Dan's house value}) + 50\% \times (\text{Dan's house value})$$
$$= 1 \times (\text{Dan's house value}) + 0.5 \times (\text{Dan's house value})$$
$$= 1.5 \times (\text{Dan's house value})$$

Note that $90 000 = 1.5 × (Dan's house value).

$$\text{Dan's house value} = \$90\ 000 \div 1.5 = \$60\ 000$$

(Note that we have used factoring in the solution to this example. The method used here is very similar to the method used to determine John's salary. You may have used your intuition to determine this answer. We have presented the math reasoning here.)

Value of Nancy's car (note that we must determine this before we can calculate the value of Dan's sailboat): The value of John's sailboat equals 30% more than the value of Nancy's car.

$$\$5200 = (\text{Nancy's car}) + 30\% \times (\text{Nancy's car})$$
$$= 1 \times (\text{Nancy's car}) + 0.3 \times (\text{Nancy's car})$$
$$= 1.3 \times (\text{Nancy's car})$$
$$\text{Nancy's car} = \$5200/1.3$$
$$= \$4000$$

The answer to this problem and to the previous problem have been determined using the same method. Does the answer seem reasonable?

a.

Person	Age	Salary	House	Boat	Car
Dan	46	$32 000	$60 000	$12 000	
John	36	$16 000	$90 000	$ 5 200	
Nancy	18	$12 000			$4 000

Exercise Set 22.4, *Cont.*

    b. Value of Dan's sailboat:

$$\text{Dan's sailboat} = 3 \times (\text{value of Nancy's car})$$
$$= 3 \times \$4000 = \$12\ 000$$

Thus,

Dan's boat value = (John's boat value + some %) × (John's boat value)

$\$12\ 000 = \$5200 + (\text{some \%})(\$5200)$

$\$12\ 000 = \$5200 \times (1 + \text{some \%})$

Divide both sides by $5200.

$$2.308 = 1 + \text{some \%}$$

Subtract 1 from both sides.

$$1.308 = \text{some \%}$$

Thus, Dan's boat is 30.8% more valuable than John's boat.

3. *Known:*

    Pennies: Diana has two; Sue has twice as many pennies as Diana. Janet has triple as many pennies as Diana.

    Nickels: Sue has twice as many nickels as she has pennies. Diana has one-half as many nickels as Sue. Janet has one more nickel than she has quarters.

    Dimes: Diana has as many dimes as Janet has pennies. Janet has one-half as many dimes as Diana. Sue has twice as many dimes as Janet has nickels.

    Quarters: Diana has one-half as many quarters as she has pennies. Janet has as many quarters as Sue has pennies. Sue has as many quarters as Janet has dimes.

Unknown:

    all values except for Diana's pennies

    number of coins that each woman has placed on the table

    amount of money that each woman has placed on the table

Math:

    multiplication, division, and addition of the numbers (Note that you must determine some values before others can be determined.)

Solution:

    Pennies: Sue has twice (2 ×) as many pennies as Diana.

$$2 \times 2 = 4$$

Janet has triple (3 ×) as many pennies as Diana.

$$3 \times 2 = 6$$

Nickels: Sue has twice (2 ×) as many nickels as she has pennies.

$$2 \times 4 = 8$$

Diana has one-half [(1/2) ×] as many nickels as Sue.

$$(1/2) \times 8 = 4$$

We can determine the number of nickels Janet has after we determine the number of quarters that Janet has.

Dimes: Diana has as many dimes as Janet has pennies (6). Janet has one-half [(1/2) ×] as many dimes as Diana.

$$(1/2) \times 6 = 3$$

We can determine the number of dimes Sue has after we determine the number of nickels that Janet has.

Quarters: Diana has one-half [(1/2) ×] as many quarters as she has pennies.

$$(1/2) \times 2 = 1$$

Janet has as many quarters as Sue has pennies (4). Sue has as many quarters as Janet has dimes (3).

Now we can say that Janet has 5 nickels, or one more nickel than she has quarters.

We can also say that Sue has twice (2 ×) as many dimes as Janet has nickels (5).

$$2 \times 5 = 10$$

a. The numbers of **coins** each woman has placed on the table are as follows.

Diana: 2 + 4 + 6 + 1 = 13
Sue: 4 + 8 + 10 + 3 = 25
Janet: 6 + 5 + 3 + 4 = 18

b. The amounts of **money** each woman has placed on the table are as follows.

Diana: (2 × 0.01 + 4 × 0.05 + 6 × 0.10 + 1 × 0.25) dollars
         = 1.07 dollars
Sue: (4 × 0.01 + 8 × 0.05 + 10 × 0.10 + 3 × 0.25) dollars
         = 2.19 dollars
Janet: (6 × 0.01 + 5 × 0.05 + 3 × 0.10 + 4 × 0.25) dollars
         = 1.61 dollars

Person	Pennies	Nickels	Dimes	Quarters	Total Number of Coins	Total Value
Diana	2	4	6	1	13	$1.07
Sue	4	8	10	3	25	$2.19
Janet	6	5	3	4	18	$1.61

Exercise Set 22.5

1. *Known:*

    Today, Jim is twice as old as John.

    Five years from now, John will be twenty.

    Ten years from now, Jack will be John's present age.

    Unknown:

    present ages of Jim, John, and Jack

    Jim's and Jack's ages five years from now

    Jim's, John's, and Jack's ages ten years from now

    the age comparison, ten years from now, of Jim to John in percent

    Math:

    multiplication, addition, and subtraction

    determining a percent

    Solution:

    John's age: Five years from now, John will be twenty. Therefore, John's present age is

$$20 - 5 = 15$$

    John's age in ten years will be

$$15 + 10 = 25$$

    Jim's age: At present, Jim is twice (2 ×) as old as John.

$$2 \times 15 = 30$$

    Therefore, five years from now, Jim's age will be

$$30 + 5 = 35$$

    and, ten years from now, Jim's age will be

$$30 + 10 = 40$$

    Jack's age: Ten years from now, Jack will be John's present age (15). Therefore, at present, Jack's age is

$$15 - 10 = 5$$

    and, five years from now, Jack's age will be

$$5 + 5 = 10$$

   a.

Person	Present Age	Age 5 Yr from Now	Age 10 Yr from Now
Jim	30	35	40
John	20 − 5 = 15	20	20 + 5 = 25
Jack	5	10	15

   b. Age ratios (in percent):

   Jim to John is 160%; John to Jim is 62.5%

2. *Known:*

    A car begins at a reference point. For two seconds, the driver accelerates the car at 4 m/s^2. The driver then stops accelerating and drives at a constant velocity for four more seconds.

  Unknown:

    the values of distance (s), velocity (v), and acceleration (a), for each of the six seconds of travel described

  Math:

    Two sets of formulas:

        the formula for acceleration, velocity, and distance when acceleration is constant

        the formulas for velocity and for distance when velocity is constant

  Solution:

    For the first two seconds (when $a = 4$ m/s^2),

$$t = 0 \text{ s}$$
$$v = (4 \text{ m/s}^2) \times (0 \text{ s}) = 0 \text{ m/s}$$
$$s = (1/2) \times (4 \text{ m/s}^2) \times (0 \text{ s})^2 = 0 \text{ m}$$

$$t = 1 \text{ s}$$
$$v = (4 \text{ m/s}^2) \times (1 \text{ s}) = 4 \text{ m/s}$$
$$s = (1/2) \times (4 \text{ m/s}^2) \times (1 \text{ s})^2 = 2 \text{ m}$$

$$t = 2 \text{ s}$$
$$v = (4 \text{ m/s}^2) \times (2 \text{ s}) = 8 \text{ m/s}$$
$$s = (1/2) \times (4 \text{ m/s}^2) \times (2 \text{ s})^2 = 8 \text{ m}$$

For the next four seconds (when $a = 0$ m/s^2, the driver is not accelerating, and $v = 8$ m/s is constant):

Apply the formula for determining the distance traveled by an object whose velocity is constant.

Note that the velocity becomes a constant at a point that is **not** the reference point. Therefore, for these next four seconds, begin counting time (a new $t = 0$) from the time when velocity becomes constant (the old $t = 2$ s).

Use the formula for distance $s = v\,\Delta t + s_0$, where s_0 is the distance of the object from a reference point when the velocity of the object becomes constant.

After 1 second of traveling at a constant velocity ($\Delta t = 1$ s),

$$s = [(8 \text{ m/s}) \times (1 \text{ s})] + 8 \text{ m}$$
$$= 8 \text{ m} + 8 \text{ m}$$
$$= 16 \text{ m}$$

After 2 seconds of traveling at a constant velocity ($\Delta t = 2$ s),

$$s = [(8 \text{ m/s}) \times (2 \text{ s})] + 8 \text{ m}$$
$$= 16 \text{ m} + 8 \text{ m}$$
$$= 24 \text{ m}$$

Exercise Set 22.5, *Cont.*

After 3 seconds of traveling at a constant velocity ($\Delta t = 3$ s),

$$s = [(8 \text{ m/s}) \times (3 \text{ s})] + 8 \text{ m}$$
$$= 24 \text{ m} + 8 \text{ m}$$
$$= 32 \text{ m}$$

After 4 seconds of traveling at a constant velocity ($\Delta t = 4$ s),

$$s = [(8 \text{ m/s}) \times (4 \text{ s})] + 8 \text{ m}$$
$$= 32 \text{ m} + 8 \text{ m}$$
$$= 40 \text{ m}$$

a. and **b.**

t	s	v	a
0	0	0	4
1	2	4	4
2	8	8	4
3	16	8	0
4	24	8	0
5	32	8	0
6	40	8	0

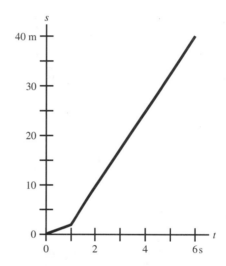

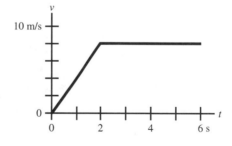

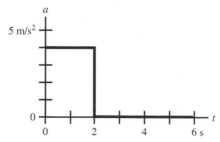

Exercise Set 22.5, *Cont.*

3. *Known:*

  Aircraft 1 leaves Brisbane; $v = 500$ mi/hr N.

  Aircraft 2 leaves Brisbane at the same time as aircraft 1;
  $v = 50$ mi/hr (N) faster than aircraft 1.

  Aircraft 3 leaves Smiles Island at the same time as aircrafts 1 and 2;
  $v = 450$ mi/hr S.

  Smiles Island is 2000 miles North of Brisbane.

Unknown:

  the distance of aircraft 1, aircraft 2, and aircraft 3 from Brisbane at one,
  two, three, and four hours after takeoff

  graph of these data

  the time when aircrafts 1 and 2 will "meet" (intersect) aircraft 3

  the time when aircrafts 1 and 2 will arrive at Smiles Island

  the time when aircraft 3 will arrive at Brisbane

Math:

  the formula for distance where velocity is constant

Solution:

Distance of aircraft 1 from Brisbane when $v = 500$ mi/hr N:

When $t = 0$ hr,

$$s = 500 \text{ mi/hr N} \times 0 \text{ hr} = 0 \text{ mi}$$

When $t = 1$ hr,

$$s = 500 \text{ mi/hr N} \times 1 \text{ hr} = 500 \text{ mi}$$

When $t = 2$ hr,

$$s = 500 \text{ mi/hr N} \times 2 \text{ hr} = 1000 \text{ mi}$$

When $t = 3$ hr,

$$s = 500 \text{ mi/hr N} \times 3 \text{ hr} = 1500 \text{ mi}$$

When $t = 4$ hr,

$$s = 500 \text{ mi/hr N} \times 4 \text{ hr} = 2000 \text{ mi}$$

Distance of aircraft 2 from Brisbane when $v = (500 + 50)$ mi/hr N
$= 550$ mi/hr N:

When $t = 0$ hr,

$$s = 550 \text{ mi/hr N} \times 0 \text{ hr} = 0 \text{ mi}$$

When $t = 1$ hr,

$$s = 550 \text{ mi/hr N} \times 1 \text{ hr} = 550 \text{ mi}$$

When $t = 2$ hr,

$$s = 550 \text{ mi/hr N} \times 2 \text{ hr} = 1100 \text{ mi}$$

When $t = 3$ hr,

$$s = 550 \text{ mi/hr N} \times 3 \text{ hr} = 1650 \text{ mi}$$

When $t = 4$ hr,

$$s = 550 \text{ mi/hr N} \times 4 \text{ hr} = 2200 \text{ mi}$$

Exercise Set 22.5, *Cont.*

Distance of aircraft 3 from Brisbane when $v = 450$ mi/hr S and the aircraft starts at Smiles Island (2000 miles N of the reference, which is Brisbane) (Recall that when distance is described as a vector, distance traveled to the south or west is negative. Velocity to the south or west is also considered to be negative):

When $t = 0$ hr,

$$s = 450 \text{ mi/hr S} \times 0 \text{ hr} + 2000 \text{ mi}$$
$$= -450 \text{ mi/hr} \times 0 \text{ hr} + 2000 \text{ mi} = 0 \text{ mi} + 2000 \text{ mi} = 2000 \text{ mi}$$

When $t = 1$ hr,

$$s = 450 \text{ mi/hr S} \times 1 \text{ hr} + 2000 \text{ mi}$$
$$= -450 \text{ mi/hr} \times 1 \text{ hr} + 2000 \text{ mi} = -450 \text{ mi} + 2000 \text{ mi} = 1550 \text{ mi}$$

When $t = 2$ hr,

$$s = 450 \text{ mi/hr S} \times 2 \text{ hr} + 2000 \text{ mi}$$
$$= -450 \text{ mi/hr} \times 2 \text{ hr} + 2000 \text{ mi} = -900 \text{ mi} + 2000 \text{ mi} = 1100 \text{ mi}$$

When $t = 3$ hr,

$$s = 450 \text{ mi/hr S} \times 3 \text{ hr} + 2000 \text{ mi}$$
$$= -450 \text{ mi/hr} \times 3 \text{ hr} + 2000 \text{ mi} = -1350 \text{ mi} + 2000 \text{ mi} = 650 \text{ mi}$$

When $t = 4$ hr,

$$s = 450 \text{ mi/hr S} \times 4 \text{ hr} + 2000 \text{ mi}$$
$$= -450 \text{ mi/hr} \times 4 \text{ hr} + 2000 \text{ mi} = -1800 \text{ mi} + 2000 \text{ mi} = 200 \text{ mi}$$

a.

Aircraft 1		Aircraft 2		Aircraft 3	
t	s	t	s	t	s
0	0	0	0	0	2000
1	500	1	550	1	1550
2	1000	2	1100	2	1100
3	1500	3	1650	3	650
4	2000	4	2200	4	200

Exercise Set 22.5, *Cont.*

b.

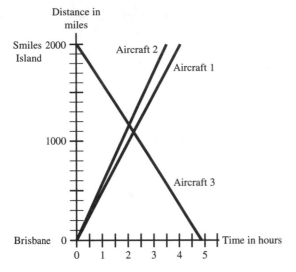

Examine the preceding chart.

c. Aircraft 1 will "meet" aircraft 3 after approximately 2.25 hr.
Aircraft 2 will "meet" aircraft 3 after 2 hr.

d. Aircraft 1 will arrive at Smiles Island after 4 hr.

e. Aircraft 2 will arrive at Smiles Island after approximately 3.5 hr.

f. Aircraft 3 will arrive at Brisbane after approximately 4.5 hr. (To determine this information, you can extend the graph of aircraft 3 because its velocity remains constant.)

CHAPTER 23

Exercise Set 23.1

1. *Unknown:*
the total number of people in all of the houses
the number of people living in the four-family apartments
the number of households in this neighborhood
the number of households of each type
the average number of people per household in each type of building
the number of people living in the new school district

Known (useful information only):
There are 39 houses in a neighborhood.
There are 38 people living in the 12 single-family houses.
There are 50 people living in the 7 two-family houses.
There are 61 people living in the 6 three-family houses.
There are 4 four-family apartments.
The average number of persons per household in this neighborhood is 3.20.

Math:
addition, subtraction, multiplication, and division
averaging

Exercise Set 23.1, *Cont.*

Solution:

a.–c. The number of households in this neighborhood will be used to determine the number of people living in the four-family apartments. Multiply the number of each type of building by the number of units in that type of building.

$$(12 \times 1 + 7 \times 2 + 6 \times 3 + 4 \times 4) \text{ households}$$
$$= (12 + 14 + 18 + 16) \text{ households} = 60 \text{ households}$$

The number of people living in the four-family apartments: Determine the total number of people in all of the houses.

Total number of households $\times$ average number of persons per household
$= 60 \text{ households} \times 3.20 \text{ people per household} = 192 \text{ people}$

The number of people living in the 4-family households:

Total number of people $-$ number of people in all of the other types of houses
$= (192) - (38 + 50 + 61) \text{ people}$
$= (192 - 149) \text{ people} = 43 \text{ people}$

The average number of people per household in the single-family houses:

$$38 \text{ people} \div 12 \text{ households} = 3.167 \text{ people per house}$$

The average number of people per household living in the two-family houses:

$$50 \text{ people} \div 14 \text{ households} = 3.571 \text{ people per household}$$

The average number of people per household living in the three-family houses:

$$61 \text{ people} \div 18 \text{ households} = 3.389 \text{ people per household}$$

The average number of people per household in the four-family houses:

$$43 \text{ people} \div 16 \text{ households} = 2.688 \text{ people per household}$$

(Note that we have divided by the number of **households** of each type of building, which is the number of each type of building multiplied by the number of units in that type of building).

d. The number of people living in the new school district:

Total number of people $-$ number of people in the single-family houses
$= (192 - 38) \text{ people} = 154 \text{ people}$

Type of Building	Number of Houses	Number of Households	Number of People
Single-family	12	12	38
Two-family	7	14	50
Three-family	6	18	61
Four-family	4	16	43

2. *Unknown:*

the number of people traveling each day of the week, in percent
the average number of people traveling per week
the average number of bags carried each week per person
the total number of bags carried in a week
the average number of bags carried each day per person

Known (useful information only):

Day of Week	Number of Passengers	Number of Bags
Sun.	270	521
Mon.	544	870
Tues.	322	708
Wed.	286	604
Thur.	310	732
Fri.	650	1235
Sat.	189	467

Math:

addition
determining percents
averaging

Solution:

a. The number of people traveling each day of the week in percent for each day is the average number of people traveling that day divided by the average number of people traveling in a week times 100. The average number of people traveling in a week is

(270 + 544 + 322 + 286 + 310 + 650 + 189) passengers/week
= 2571 passengers/week

Therefore, the numbers of people traveling each day, in percent, are as follows.
Sunday:

$[(270 \div 2571)]$ passengers $\times$ 100
= 10.502% of the passengers

Monday:

$[(544 \div 2571)]$ passengers $\times$ 100
= 21.16% of the passengers

Tuesday:

$[(322 \div 2571)]$ passengers $\times$ 100
= 12.52% of the passengers

Wednesday:

$[(286 \div 2571)]$ passengers $\times$ 100
= 11.124% of the passengers

Exercise Set 23.1, *Cont.*

Thursday:

$$[(310 \div 2571)] \text{ passengers} \times 100$$
$$= 12.058\% \text{ of the passengers}$$

Friday:

$$[(650 \div 2571)] \text{ passengers} \times 100$$
$$= 25.28\% \text{ of the passengers}$$

Saturday:

$$[(189 \div 2571)] \text{ passengers} \times 100$$
$$= 7.351\% \text{ of the passengers}$$

Note: The total of the percents is 99.995%.

b. The average number of bags carried each week per person is

Total number of bags ÷ average number of passengers per week
$$= (521 + 870 + 708 + 604 + 732 + 1235 + 467) \text{ bags}$$
$$\div 2571 \text{ passengers per week}$$
$$= 1.998 \text{ bags per passenger per week}$$

The average number of bags carried each day of the week per person is

Number of bags carried that day ÷ average number of passengers that day

Sunday:

$$(521 \text{ bags}) \div (270 \text{ passengers})$$
$$= 1.9296 \text{ bags per passenger}$$

Monday:

$$(870 \text{ bags}) \div (544 \text{ passengers})$$
$$= 1.5993 \text{ bags per passenger}$$

Tuesday:

$$(708 \text{ bags}) \div (322 \text{ passengers})$$
$$= 2.199 \text{ bags per passenger}$$

Wednesday:

$$(604 \text{ bags}) \div (286 \text{ passengers})$$
$$= 2.112 \text{ bags per passenger}$$

Thursday:

$$(732 \text{ bags}) \div (310 \text{ passengers})$$
$$= 2.361 \text{ bags per passenger}$$

Friday:

$$(1235 \text{ bags}) \div (650 \text{ passengers})$$
$$= 1.900 \text{ bags per passenger}$$

Saturday:

$$(467 \text{ bags}) \div (189 \text{ passengers})$$
$$= 2.470 \text{ bags per passenger}$$

Exercise Set 23.2

1. *Unknown:*
    the price per ounce for each brand of cheese
    the brand of cheese with the lowest price per ounce
    the lowest-priced bar, cube, and wedge of cheese
   Known (useful information only):
    the weight (in ounces or grams) of each brand of cheese
    the package shape of each brand of cheese
    the price of each brand of cheese
   Math:
    use of the unity method to convert units
    division
   Solution:
   a. To determine the price per ounce for each brand of cheese, divide the total cost of each cheese by the number of ounces. Convert grams to ounces where necessary.
   Jason's:

   $$(2.19 \text{ dollars}) \div (12 \text{ ounces}) = 0.18 \text{ dollar/ounce}$$

   Nathaniel's:

   $$(1.39 \text{ dollars}) \div (8 \text{ ounces}) = 0.17 \text{ dollar/ounce}$$

   Lemieux's:

   $$[(1.69 \text{ dollars}) \div (252 \text{ grams}) \times (28 \text{ grams/1 ounce})]$$
   $$= (1.69 \times 28) \div (252) \text{ dollars/ounce} = 0.19 \text{ dollar/ounce}$$

   Best Buy:

   $$(2.19 \text{ dollars}) \div (14 \text{ ounces}) = 0.16 \text{ dollar/ounce}$$

   Lauderleif:

   $$[(1.19 \text{ dollars}) \div (196 \text{ grams})] \times (28 \text{ grams/1 ounce})$$
   $$= (1.19 \times 28) \div (196) \text{ dollars/ounce} = 0.17 \text{ dollar/ounce}$$

   Old Larder:

   $$(1.49 \text{ dollars}) \div (10 \text{ ounces}) = 0.15 \text{ dollar/ounce}$$

   b. The brand of cheese with the lowest price is Old Larder at 0.15 dollar per ounce.
   c. The lowest-priced bar of cheese is Lauderleif, the only bar-shaped cheese.
    The lowest-priced cube of cheese is Jason's at 0.18 dollar/ounce.
    The lowest-priced wedge of cheese is Old Larder at 0.15 dollar/ounce.

2. *Unknown:*
    the grandchildren of Alfredo and Marguarita
    the grandchildren of Benjamin and Dorris
   Known (useful information only):
    Alfredo and Marguarita's children are Jose, Sammy, and Alberto.
    Alberto and Barbara's children are Albert, Susan, and Maria.

Exercise Set 23.2, *Cont.*

Benjamin and Dorris's children are William, Anita, Carlotta, Delores, and Ricardo.

Jose and Anita's children are Carlos, Nancy, and Roberto.

Pierre and Carlotta's children are Consuelo, Billy, Anatole, Jeanette, and Nellie.

The sketch is as follows:

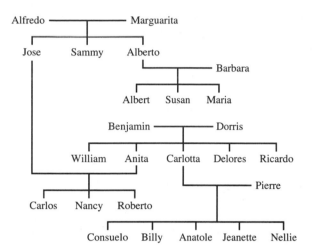

Solution:

a. The grandchildren of Alfredo and Marguarita are Carlos, Nancy, and Roberto; Albert, Susan, and Maria.

b. The grandchildren of Benjamin and Dorris are Carlos, Nancy, and Roberto; Consuelo, Billy, Anatole, Jeannette, and Nellie.

Exercise Set 23.3

1. *Unknown:*

the amount of shelf space (width) that each person will require

the average weight per book for each person

the average weight per book for the entire family

the average age of the persons in this household

the person whose books are the heaviest

the person whose books require the most space

the number of shelves that may be constructed vertically (The height of the set of bookshelves is to be as close to 4'6" as possible, but no higher, and every shelf must be able to contain the tallest book of any member.)

Known (useful information only):

each family member's age

the height of their tallest book

the average width of their books

the number of their books

the weight of their books

the width of each board

Exercise Set 23.3, *Cont.*

Math:

averaging

addition and multiplication

Solution:

a. The amount of shelf space that each person will require is (the average width per book) × (the number of books) for each person.

Mom: (1.2 in./book) × (30 books) = 36 in.

Dad: (1.6 in./book) × (40 books) = 64 in.

Billy: (1.3 in./book) × (30 books) = 39 in.

Nancy: (1.5 in./book) × (34 books) = 51 in.

Norma: (0.8 in./book) × (25 books) = 20 in.

Craig: (0.5 in./book) × (16 books) = 8 in.

b. The average weight per book for each person is (the weight of each person's books) ÷ (the total number of their books).

Mom: (72 lb) ÷ (30 books) = 2.4 lb/book

Dad: (149 lb) ÷ (40 books) = 3.725 lb/book

Billy: (80 lb) ÷ (30 books) = 2.667 lb/book

Nancy: (104 lb) ÷ (34 books) = 3.059 lb/book

Norma: (47 lb) ÷ (25 books) = 1.88 lb/book

Craig: (14 lb) ÷ (16 books) = 0.875 lb/book

The average weight per book for the entire family is

Total weight of every person's books ÷ total number of books

= (72 + 149 + 80 + 104 + 47 + 14) lb

   ÷ (30 + 40 + 30 + 34 + 25 + 16) books

= (466 lb) ÷ (175 books) = 2.663 lb/book

c. The average age of the persons in the household is

(39 + 42 + 14 + 11 + 10 + 8) years ÷ (6 persons)

= (124 years) ÷ (6 persons) = 20.67 years/person

d. The person whose books are the heaviest is Dad (149 lb). This is determined directly from the information in part (b).

e. The person whose books will require the most space is Dad [see the information in part (a)].

f. The number of shelves that may be constructed vertically: Each shelf must be able to contain the tallest book of any family member. The tallest book of any family member is 14 in. This may be determined from the chart that appears in the original exercise.

Add shelf thickness (3/4 in.) to the space for the tallest book (14 in.). Note that

(0.75 + 14 + 0.75 + 14 + 0.75 + 14) in.

= 44.25 in. or $44\frac{1}{4}$ in. or 3 ft $8\frac{1}{4}$ in.

Therefore, three shelves is the greatest number that can be constructed vertically without being taller than 4 ft 6 in. or 54 in.

2. *Unknown:*

the total payroll for this company

the extra amount of payroll contribution for Social Security (7%) that the company must add to its payroll

the average hours worked per person per week

the average hours worked per person per day

Known (useful information only):

Natalie (the owner) works 54 hours and earns a total of $550 per week.

Jack works 40 hours per week and earns $7.52 per hour.

Mary works 6 hours per day for 4 days and earns $6.84 per hour.

Al works 18 hours plus 20 hours this week and is paid $5.66 per hour.

Jan works 40 hours per week and 6 unpaid hours this week; she is paid $8.66 per hour.

Bob works 30 hours this week and is paid $9.24 per hour.

Fred works 40 hours per week and is paid $4.50 per hour.

Betty works 40 hours per week and is paid $4.50 per hour.

Math:

multiplying rates

averaging

use of the unity method to convert units

Solution:

a. The total payroll for the company: Determine each person's salary per week.

Natalie: She earns $550/wk.

Jack: (40 hr/wk) $\times$ $7.52/hr = $300.80/wk

Mary: (6 hr/day) $\times$ (4 days/wk) $\times$ ($6.84/hr) = 164.16 dollars/wk

Al: (18 + 20) hrs/wk $\times$ $5.66/hr = $215.08/wk

Jan: (40 hr/wk) $\times$ ($8.66/hr) = $346.40/wk

Bob: (30 hr/wk) $\times$ ($9.24/hr) = $277.20/wk

Fred: (40 hr/wk) $\times$ ($4.50/hr) = $180.00/wk

Betty: (40 hr/wk) $\times$ ($4.50/hr) = $180.00/wk

Combine everyone's salaries to determine the total payroll.

$$(550 + 300.80 + 164.16 + 215.08 + 346.40$$
$$+ 277.20 + 180.00 + 180.00) \text{ dollars/week}$$
$$= \$2213.64/\text{week}$$

b. The extra amount of payroll contribution (7%) for Social Security is

$$(\$2213.64/\text{week}) \times 7/100$$
$$= \$2213.64/\text{week} \times 0.07$$
$$= \$154.95/\text{week}$$

Exercise Set 23.3, *Cont.*

  c. The average hours worked per week is

  Total number of hours worked per week ÷ number of persons
  = (54 + 40 + 24 + 38 + 46 + 30 + 40 + 40) hours/week ÷ 8 persons
  = (312 hr/wk) ÷ 8 persons
  = 39 hr/wk per person, or
  = 39 hr/person per week

   Note that Jan worked 46 hours this week, but she was paid for only
   40 hours.

  d. The average number of hours worked per person per day is

$$(39 \text{ hr/person per wk}) \times (1 \text{ wk/5 days})$$
$$= 7.8 \text{ hr/person per day or}$$
$$= 7.8 \text{ hr/day per person}$$

Exercise Set 23.4

1. *Unknown:*
   percent of the following types of buttons:
     red, blue, black, and 4-cm-diameter
   after 12 packages of six 5-cm green buttons, eight 4-cm black buttons,
   and four 3-cm red buttons have been prepared, the number of these three
   types of buttons, as well as the 3-cm green and 3-cm black buttons, that
   remain
   the percent of each of the eleven types of buttons, as well as the total
   number of buttons, now remaining
   the number of packages of five 3-cm red buttons, three 3-cm green buttons, and seven 3-cm black buttons that can be assembled from the remaining inventory
   the number of each of these three types of buttons that remain after the
   packages have been assembled
Known (useful information only):
   the number of each of the eleven types of buttons
Math:
   addition and subtraction
   percents
Solution:
a. and **b.** Percent of red, blue, black, and 4-cm-diameter buttons: In order
         to determine these percents, first determine the total number of
         buttons.

(66 + 142 + 85 + 210 + 66 + 93 + 80 + 211 + 96 + 64 + 99) buttons
= 1212 buttons

   The percent of each type of red, blue, black, and 4-cm buttons is

Number of one type of button ÷ total number of buttons × 100

   Percent of red buttons:

$$[(66 + 96) \div 1212] \times 100 \text{ red buttons} = 13.37\% \text{ red}$$

Exercise Set 23.4, *Cont.*

Percent of blue buttons:

$$[210 \div 1212] \times 100 \text{ blue buttons} = 17.33\% \text{ blue}$$

Percent of black buttons:

$$[(93 + 211 + 99) \div 1212] \times 100 \text{ black buttons} = 33.25\% \text{ black}$$

Percent of 4-cm-diameter buttons:

$$[(80 + 211) \div 1212] \times 100 \text{ 4-cm-diameter buttons} = 24.01\%$$

c. and **d.** The numbers of 5-cm green buttons, 4-cm black buttons, and 3-cm red buttons after the 12 packages have been assembled: Determine how many of each type of button will be removed, and subtract this quantity from the initial quantity.

5-cm green buttons:

$$(12 \times 6) \text{ 5-cm green buttons}$$
$$= 72 \text{ each of 5-cm green buttons}$$

The number of buttons that remain are

$$(85 - 72) \text{ 5-cm green buttons}$$
$$= 13 \text{ each of 5-cm green buttons}$$

4-cm black buttons:

$$(12 \times 8) \text{ 4-cm black buttons}$$
$$= 96 \text{ each of 4-cm black buttons}$$

The number of buttons that remain are

$$(211 - 96) \text{ 4-cm black buttons}$$
$$= 115 \text{ each of 4-cm black buttons}$$

3-cm red buttons:

$$(12 \times 4) \text{ 3-cm red buttons}$$
$$= 48 \text{ each of 3-cm red buttons}$$

The number of buttons that remain are

$$(96 - 48) \text{ 3-cm red buttons}$$
$$= 48 \text{ each of 3-cm red buttons}$$

Percent of each of the eleven types of buttons that remain: In order to determine these percents, first determine the remaining total number of buttons.

$$(66 + 142 + 13 + 210 + 66 + 93 + 80 + 115 + 48 + 64 + 99) \text{ buttons}$$
$$= 996 \text{ buttons}$$

Percent of 5-cm red buttons:

$$[(66 \div 996)] \times 100 \text{ 5-cm red buttons}$$
$$= 6.63\% \text{ 5-cm red buttons}$$

Percent of 5-cm orange buttons:

$$[(142 \div 996)] \times 100 \text{ 5-cm buttons}$$
$$= 14.26\% \text{ 5-cm orange buttons}$$

Percent of 5-cm green buttons:

$$[(13 \div 996)] \times 100 \text{ 5-cm green buttons}$$
$$= 1.31\% \text{ 5-cm green buttons}$$

Percent of 5-cm blue buttons:

$$[(210 \div 996)] \times 100 \text{ 5-cm blue buttons}$$
$$= 21.08\% \text{ 5-cm blue buttons}$$

Percent of 3-cm purple buttons:

$$[(66 \div 996)] \times 100 \text{ 3-cm purple buttons} = 6.63\%$$

Percent of 5-cm black buttons:

$$[(93 \div 996)] \times 100 \text{ 5-cm black buttons} = 9.34\%$$

Percent of 4-cm orange buttons:

$$[(80 \div 996)] \times 100 \text{ 4-cm orange buttons} = 8.03\%$$

Percent of 4-cm black buttons:

$$[(115 \div 996)] \times 100 \text{ 4-cm black buttons} = 11.55\%$$

Percent of 3-cm red buttons:

$$[(48 \div 996)] \times 100 \text{ 3-cm red buttons} = 4.82\%$$

Percent of 3-cm green buttons:

$$[(64 \div 996)] \times 100 \text{ 3-cm green buttons} = 6.43\%$$

Percent of 3-cm black buttons:

$$[(99 \div 996)] \times 100 \text{ 3-cm black buttons} = 9.94\%$$

e. The number of packages of five 3-cm red, three 3-cm green, and seven 3-cm black buttons that can now be assembled: Indicate the number of each type of button that may be removed.

There are forty-eight 3-cm red buttons.

$$48 \div 5 = 9 \quad \text{with a remainder of 3}$$

There are sixty-four 3-cm green buttons.

$$64 \div 3 = 21 \quad \text{with a remainder of 1}$$

There are ninety-nine 3-cm black buttons.

$$99 \div 7 = 14 \quad \text{with a remainder of 1}$$

Therefore, no more than nine packages of five 3-cm red, three 3-cm green, and seven 3-cm black buttons may be assembled.

Exercise Set 23.4, *Cont.*

    f. The quantity of these three types of buttons that remain in inventory after the nine packages have been assembled are as follows.
        3-cm red buttons:

$$48 - (9 \times 5) \text{ 3-cm red buttons}$$
$$= 3 \text{ each of 3-cm red buttons}$$

        3-cm green buttons:

$$64 - (9 \times 3) \text{ 3-cm green buttons}$$
$$= 37 \text{ each of 3-cm green buttons}$$

        3-cm black buttons:

$$99 - (9 \times 7) \text{ 3-cm black buttons}$$
$$= 36 \text{ each of 3-cm black buttons}$$

2. *Unknown:*

    the volume of the ingredients
    the size of the pot (one-, two-, three-, or five-quart) that should be chosen if the pot is to be no more than one-half full of all of the ingredients
    the ratio of dry to wet ingredients in the recipe

Known (useful information only):

    Ingredients in the recipe:
        3 cups of flour
        2 tablespoons of salt
        1/2 pound and 1 cup of water
        8 tablespoons of sugar

Math:

    use of the unity method to convert units
    division to determine ratio

Solution:

    a. The volume of ingredients: In order for us to combine the four ingredients, they must have the same units of measure. (In this solution, they have been converted to quarts because the pots to be selected are given as quart sizes.)

        Flour:

$$(3 \text{ C}) \times (1 \text{ pt/2 C}) \times (1 \text{ qt/2 pt})$$
$$= [3/(2 \times 2)] \text{ qt} = 0.75 \text{ qt}$$

        Salt:

$$(2 \text{ T}) \times (1 \text{ pt/32 T}) \times (1 \text{ qt/2 pt})$$
$$= [2/(32 \times 2)] \text{ qt} = 0.03125 \text{ qt}$$

        Water:

$$(1/2 \text{ lb}) \times (16 \text{ oz/1 lb}) \times (1 \text{ qt/32 oz})$$
$$= [16/(2 \times 32)] \text{ qt} = 0.25 \text{ qt}$$

        Water again:

$$(1 \text{ C}) \times (1 \text{ pt/2 C}) \times (1 \text{ qt/2 pt})$$
$$= [1/(2 \times 2)] \text{ qt} = 0.25 \text{ qt}$$

Exercise Set 23.4, *Cont.*

Sugar:

$$(8 \text{ T}) \times (1 \text{ pt}/32 \text{ T}) \times (1 \text{ qt}/2 \text{ pt})$$
$$= [8/(32 \times 2)] \text{ qt} = 0.125 \text{ qt}$$

The total volume of all ingredients is

$$(0.75 + 0.03125 + 0.25 + 0.25 + 0.125) \text{ qt}$$
$$= 1.40625 \text{ qt}$$

The size of the pot that should be chosen: The pot is to be no more than one-half full. Therefore, the 3-qt pot should be chosen because

$$(1/2 \times 3) \text{ qt} = 1.5 \text{ qt}$$

Note that 1.5 quarts is slightly larger than the 1.40625 quarts of ingredients.

How full is the pot? Note the ratio

$$(1.40625 \text{ qt}) \div (3 \text{ qt}) = 0.4688 \quad \text{or} \quad 46.88\% \text{ full}$$

The pot is slightly less than one-half full.
Note that the 2-qt pot is too small because

$$(1/2 \times 2 \text{ qt}) = 1 \text{ qt}$$

which is smaller than the volume of the ingredients.
How full would the 2-quart pot be? Note the ratio

$$(1.40625 \text{ qt}) \div (2 \text{ qt})$$
$$= 0.7031, \text{ or } 70.31\% \text{ full}$$

The pot would be more than one-half full.

b. The ratio of dry to wet ingredients: Dry ingredients are flour, salt, and sugar.

$$(0.75 + 0.03125 + 0.125) \text{ qt} = 0.90625 \text{ qt}$$

Wet ingredients are water (used twice).

$$(0.25 + 0.25) \text{ qt}$$
$$= 0.50 \text{ qt}$$

The ratio of dry to wet ingredients is

$$(0.90625 \text{ qt}) \div (0.5 \text{ qt}) = 1.8125$$

Exercise Set 23.5

1. *Unknown:*
    the height of the 3-story building
    the distance from the ground to the top of the second floor
    the distance from the top of the second floor to the top of the building
   Known (useful information only):
    the distance between the surveyor and the building (40 ft)
    the angle from the ground to the top of the second floor (46.32°)
    the angle from the ground to the top of the third floor (57.68°)

Exercise Set 23.5, *Cont.*

Math:

  one of the trig formulas
  subtraction

Solution:

a. To determine the height of the 3-story building: Note that the value of the angle (57.68°) from the ground to the top of the third floor, and the adjacent side (40 ft) of this right triangle, are known. The length of the opposite side is the unknown. Therefore, use the formulas

$$\text{Opposite} = (\text{adjacent}) \times (\tan \theta)$$
$$\text{Height} = (40 \text{ ft}) \times (\tan 57.68°)$$
$$= (40 \text{ ft}) \times (1.5806)$$
$$= 63.22 \text{ ft}$$

b. To determine the distance from the top of the second floor to the top of the building: First, determine the distance from the ground to the top of the second floor. (Note that the same trig formula may be used.)

$$\text{Opposite} = (\text{adjacent}) \times (\tan \theta)$$
$$\text{Height} = (40 \text{ ft}) \times (\tan 46.32°)$$
$$= (40 \text{ ft}) \times (1.0472)$$
$$= 41.89 \text{ ft}$$

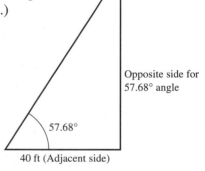

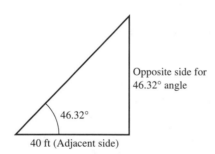

Therefore, the distance from the top of the second floor to the top of the building is

Height of building − distance from ground to top of second floor
= (63.22 − 41.89) ft
= 21.33 ft

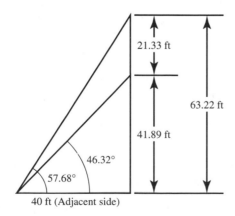

Exercise Set 23.5, *Cont.*

2. *Unknown:*

the amount of money withdrawn during June, July, and August

the amount of money deposited on the three indicated days

the difference between the total amount of money withdrawn and the total amount of money deposited

Known (useful information only):

the deposit and withdrawal activity for each of the three accounts in June, July, and August

Math:

addition and subtraction

Solution:

a. The amount of money withdrawn in June from the three accounts is

$$(4000 + 1000) \text{ dollars}$$
$$= 5000 \text{ dollars}$$

The amount of money withdrawn in July from the three accounts is

$$(500 + 1000 + 2000) \text{ dollars}$$
$$= 3500 \text{ dollars}$$

The amount of money withdrawn in August from the three accounts is

$$(1500 + 2000 + 1000) \text{ dollars}$$
$$= 4500 \text{ dollars}$$

b. The amount of money deposited on June 2 into the three accounts is

$$(2000 + 2000) \text{ dollars}$$
$$= 4000 \text{ dollars}$$

The amount of money deposited on July 4 into the three accounts is

$$(1000 + 1000) \text{ dollars}$$
$$= 2000 \text{ dollars}$$

The amount of money deposited on August 13 into the three accounts is

$$3000 \text{ dollars}$$

The difference between the total amount of money withdrawn and the total amount of money deposited is

$$[(5000 + 3500 + 4500) - (4000 + 2000 + 3000)] \text{ dollars}$$
$$= (13\ 000 - 9000) \text{ dollars}$$
$$= 4000 \text{ dollars}$$

CHAPTER 24

Exercise Set 24.1

1. The terms are $3x$, $-4yz$, and $+2xyz$; the coefficients are 3, -4, and 2.

2. The term is $5xy \div xz$; the coefficient is 5.

3. The terms are $7nmp \div q$, $-4nm$, and $9mp$; the coefficients are 7, -4, and 9.

Exercise Set 24.1, *Cont.*

4. The terms are $abc \div k$, $3bcd$, $-2ac$, and $5ad$; the coefficients are 1, 3, -2, and 5.

5. The terms are $4akz$, $-5mp/y$, and $5bdv$; the coefficients are 4, -5, and 5.

Exercise Set 24.2

1. $\underline{7xy} + \underline{3yz} - \underline{2xy}$; $5xy + 3yz$
2. $\underline{9ab} - \underline{2bc} + \underline{4ab}$; $13ab - 2bc$
3. $\underline{4mp} - \underline{7np} + \underline{3mn} - \underline{2np}$; $3mn + 4mp - 9np$
4. $\underline{5xyz} + \underline{3wxy} - \underline{2xyz} + \underline{4wxy}$; $3xyz + 7wxy$
5. $\underline{3abcd} - \underline{4bcd} + \underline{5} - \underline{2abcd} + \underline{7bcd}$; $abcd + 3bcd + 5$
   (The coefficient of the first term is 1.)
6. $\underline{5mnpq} - \underline{3mnpq} + \underline{7mnpq} + \underline{6mnpq}$; $15mnpq$
7. $\underline{7rst} - \underline{5rt} + \underline{4rs} - \underline{3rt} + \underline{5rst}$; $12rst + 4rs - 8rt$
8. $-\underline{9kmp} + \underline{8mp} + \underline{kp} - \underline{2kmp} - \underline{kp}$; $-11kmp + 8mp$
   (The kp terms cancel.)
9. $\underline{4xyz} - \underline{2wxy} + \underline{3zyx} - \underline{5xyw}$; $7xyz - 7wxy$
10. $\underline{6rs} - \underline{4ts} + \underline{11sr} + \underline{2sr} - \underline{8rst}$; $8rs + 7st - 8rst$

Exercise Set 24.3

1. $x = 72/6 = 12$
2. $x = -48/8 = -6$
3. $x = 52/-4 = -13$
4. $5x = 105$
   $x = 21$
5. $5x = 75$
   $x = 15$
6. $14x = 84$
   $x = 6$
7. $9x = 135$
   $x = 15$
8. $18x = 54$
   $x = 3$

Exercise Set 24.4

1. $7x = 12 + 4x$

   $7x - 4x = 12 + 4x - 4x$ Subtract $4x$ from both sides to gather like terms.

   $3x = 12$ Combine like terms.

   $x = 4$ Divide both sides by the coefficient of x, which is 3.

   $7(4) = 12 + 4(4)$?   Verify your result.

   $28 = 12 + 16$?

   $28 \equiv 28$ True!

Exercise Set 24.4, *Cont.*

2. $6x - 7 = 11$
    $6x - 7 + 7 = 11 + 7$ Add 7 to both sides to gather like terms.
    $6x = 18$ Combine like terms.
    $x = 3$ Divide both sides by the coefficient of x, which is 6.
    $6(3) - 7 = 11$?   Verify your result.
    $18 - 7 = 11$?
    $11 \equiv 11$ True!

3. $2x + 5 = 6x + 29$
    $2x + 5 - 6x = 6x + 29 - 6x$ Subtract $6x$ from both sides to combine like terms.
    $-4x + 5 = 29$ Combine like terms.
    $-4x + 5 - 5 = 29 - 5$ Subtract 5 from both sides to combine like terms.
    $-4x = 24$ Combine like terms.
    $x = -6$ Divide both sides by the coefficient of x, which is (-4).

    $2(-6) + 5 = 6(-6) + 29$?   Verify your result.
    $-12 + 5 = -36 + 29$?
    $-7 \equiv -7$ True!

4. $8x - 6 = 3x + 9$
    $8x - 6 - 3x = 3x + 9 - 3x$ Subtract $3x$ from both sides to combine like terms.
    $5x - 6 + 6 = 9 + 6$ Add 6 to both sides to combine like terms.
    $5x = 15$ Combine like terms.
    $x = 3$ Divide both sides by the coefficient of x, which is 5.
    $8(3) - 6 = 3(3) + 9$?   Verify your result.
    $24 - 6 = 9 + 9$?
    $18 \equiv 18$ True!

5. $3s + 8 = 5s - 10$
    $3s + 8 - 5s = 5s - 10 - 5s$ Subtract $5s$ from both sides to combine like terms.
    $-2s + 8 - 8 = -10 - 8$ Subtract 8 from both sides to combine like terms.
    $-2s = -18$ Combine like terms.
    $s = 9$ Divide both sides by the coefficient of s, which is (-2).
    $3(9) + 8 = 5(9) - 10$?   Verify your result.
    $27 + 8 = 45 - 10$?
    $35 \equiv 35$ True!

Exercise Set 24.4, *Cont.*

6. $5v - 17 = 9v + 12 - 6v$

$5v - 17 - 9v + 6v = 9v + 12 - 6v - 9v + 6v$ Force the like terms with v to the left side.

$2v - 17 + 17 = 12 + 17$ Add 17 to both sides to combine like terms.

$2v = 29$ Combine like terms.

$v = 29/2$ or 14.5   Divide both sides by the coefficient of v, which is 2.

$5(14.5) - 17 = 9(14.5) + 12 - 6(14.5)?$ Verify your result.

$72.5 - 17 = 130.5 + 12 - 87?$

$55.5 \equiv 55.5$ True!

7. $15 + 6y = 8 + 2y - 7$

$15 + 6y - 2y = 8 + 2y - 7 - 2y$ Subtract $2y$ from both sides to combine like terms.

$15 + 4y - 15 = 8 - 7 - 15$ Subtract 15 from both sides to combine like terms.

$4y = -14$ Combine like terms.

$y = -7/2$ or -3.5 Divide by the coefficient of y, which is 4.

$15 + 6(-3.5) = 8 + 2(-3.5) - 7?$ Verify your result.

$15 - 21 = 8 - 7 - 7?$

$-6 \equiv -6$ True!

8. $18t - 24 = 14t + 28$

$18t - 24 - 14t = 14t + 28 - 14t$ Subtract $14t$ from both sides to combine like terms.

$4t - 24 + 24 = 28 + 24$ Add 24 to both sides to combine like terms.

$4t = 52$ Combine like terms.

$t = 13$ Divide both sides by the coefficient of t, which is 4.

$18(13) - 24 = 14(13) + 28?$ Verify your result.

$234 - 24 = 182 + 28?$

$210 \equiv 210$ True!

9. $9Q - 44 = 16Q - 17$

$9Q - 44 - 16Q = 16Q - 17 - 16Q$ Subtract $16Q$ from both sides to combine like terms.

$-7Q - 44 + 44 = -17 + 44$ Add 44 to both sides to combine like terms.

$-7Q = 27$ Combine like terms.

$Q = -27/7$ Divide both sides by the coefficient of Q, which is (-7).

$$9\left(-\frac{27}{7}\right) - 44 = 16\left(-\frac{27}{7}\right) - 17?$$ Verify your result.

$$-\frac{243}{7} - \frac{308}{7} = -\frac{432}{7} - \frac{119}{7}?$$

$$-\frac{551}{7} \equiv -\frac{551}{7}$$ True!

Exercise Set 24.5

1. $17p - 24p = 18 - 4$
   $-7p = 14$
   $p = -2$
   $17(-2) - 18 = 24(-2) - 4?$ Verify your result.
   $-34 - 18 = -48 - 4?$
   $-52 \equiv -52$ True!

2. $85V - 49V = 54 - 72$
   $36V = -18$
   $V = -1/2$ or -0.5
   $85(-0.5) + 72 = 54 + 49(-0.5)?$ Verify your result.
   $-42.5 + 72 = 54 - 24.5?$
   $29.5 \equiv 29.5$ True!

3. $18z - 2z - 25z = -52 + 19 + 15$
   $-9z = -18$
   $z = 2$
   $18(2) - 2(2) + 52 = 19 + 25(2) + 15?$ Verify your result.
   $36 - 4 + 52 = 19 + 50 + 15?$
   $84 \equiv 84$ True!

4. $48Y + 88Y = -65 + 15 + 55 - 22$
   $136Y = -17$
   $Y = -17/136$ or -0.125 (which is $-1/8$)
   $65 + 48\left(-\frac{1}{8}\right) - 15 = 55 - 88\left(-\frac{1}{8}\right) - 22?$ Verify your result.
   $65 - 6 - 15 = 55 + 11 - 22?$
   $44 \equiv 44$ True!

5. $88 - 14u = 72u - 84$
   $-14u - 72u = -88 - 84$
   $-86u = -172$
   $u = 2$
   $72 - 14(2) + 16 = 54(2) - 84 + 18(2)?$ Verify your result.
   $72 - 28 + 16 = 108 - 84 + 36?$
   $60 \equiv 60$ True!

Exercise Set 24.5, *Cont.*

6. $17.6x - 3.5x = 14.2 + 28.1$
$14.1x = 42.3$
$x = 3$
$17.6(3) - 14.2 = 3.5(3) + 28.1$?   Verify your result.
$52.8 - 14.2 = 10.5 + 28.1$?
$38.6 \equiv 38.6$ True!

7. $-153.9 - 44.4R = 8.9R + 59.3$
$-44.4R - 8.9R = 153.9 + 59.3$
$-53.3R = 213.2$
$R = -4$
$-71.2 - 44.4(-4) - 82.7 = 26.5(-4) + 59.3 - 17.6(-4)$?   Verify your result.

$-71.2 + 177.6 - 82.7 = -106 + 59.3 + 70.4$?
$23.7 \equiv 23.7$ True!

8. First, clear of fractions; LCD = 2.
$3x - 10 = 18x + 5$
$3x - 18x = 5 + 10$
$-15x = 15$
$x = -1$
$\frac{3}{2}(-1) - 5 = 9(-1) + \frac{5}{2}$?   Verify your result.

$-\frac{3}{2} - \frac{10}{2} = -\frac{18}{2} + \frac{5}{2}$?

$-\frac{13}{2} \equiv -\frac{13}{2}$ True!

Exercise Set 24.6

1. $s = vt$; solve for v
$v = s/t = 400$ mi east/8 hr
$v = 50$ mi/hr east
400 mi east $= 50$ mi/hr east $\times$ 8 hr?   Verify your result.
400 mi east $\equiv 400$ mi east   True!

2. $s = vt$; solve for t
$t = s/v$
$t = 500$ m/(4 m/s)
$t = 125$ s
(The "meters" cancel.)
500 m $= (4$ m/s$)(125$ s$)$?   Verify your result.
500 m $\equiv 500$ m   True!

3. $A = hw$; solve for h
$h = A/w$
$h = 420$ m^2/70 m
$h = 6$ m
420 m$^2 = 6$ m $\times$ 70 m?   Verify your result.
420 m$^2 \equiv 420$ m^2 True!

4. $A = s^2$; solve for s
   $s = A^{1/2}$
   $s = (6400 \text{ m}^2)^{1/2}$
   $s = 80 \text{ m}$
   $6400 \text{ m}^2 = (80 \text{ m})^2$?   Verify your result.
   $6400 \text{ m}^2 \equiv 6400 \text{ m}^2$ True!

Answers to Chapter Tests, Part Review Tests, and Book 2 Test

Each problem in the Chapter Tests and the Part Review Tests is worth **one point** except in Chapter 15, which has only 10 problems rather than 20. Problems in Chapter 15 are worth **two points** each.

Chapter Tests and Part Review Tests Scoring

20 points maximum score
19–20   Excellent
17–18   Good
15–16   Fair
13–14   Poor
* 0–12   Reread Appendices D and E;*
*   reread the appropriate chapter(s);*
*   then take the test again.*

The Book Test scoring is given on page B–15.

Chapter 15   More Fractions

Problems are worth 2 points each.

1.	14/3	**2.**	3/56
3.	27/25	**4.**	−2/7
5.	3/35	**6.**	17/28
7.	25/22	**8.**	25/3
9.	25/4	**10.**	25/44

Chapter 16   Exponents and Radicals

1.	2^5	**2.**	$3^2 \times 5^4$
3.	$2^3/3^4$	**4.**	144
5.	$5^3 = 125$	**6.**	72
7.	1	**8.**	17
9.	$2^6 = 64$	**10.**	$1/64 = 0.015\ 625$

Chapter 16   Exponents and Radicals, *Cont.*

11. $64/(19\ 683) = 0.003\ 251\ 5$ **12.** 5×10^4

13. 2×10^{-3} **14.** 600

15. 0.4 **16.** $2^{2/2} \times 3^{2/2} = 6$

17. $2^{3/3} = 2$ **18.** $2^{6/3} = 4$

19. $+5$ or -5 **20.** -5 or $+5$

Chapter 17   The Three Dimensions

1. perimeter   **2.** area

3. 12.4 cm   **4.** $9.61\ \text{cm}^2$

5. 20.6 cm   **6.** $25.62\ \text{cm}^2$

7. 10.2 in.   **8.** $3.708\ \text{in.}^2$

9. 26.4 cm   **10.** $24.9\ \text{cm}^2$

11. 11.2π cm $= 35.186$ cm   **12.** $86.49\pi\ \text{cm}^2$ or $271.72\ \text{cm}^2$

13. $105.84\ \text{in.}^2$ **14.** $74.088\ \text{in.}^3$

15. $186.76\ \text{cm}^2$ **16.** $156.58\ \text{cm}^3$

17. $84.64\pi\ \text{in.}^2$ or $265.90\ \text{in.}^2$ **18.** $129.78\pi\ \text{in.}^3$ or $407.72\ \text{in.}^3$

19. $94.72\pi\ \text{cm}^2$ or $297.57\ \text{cm}^2$ **20.** $124.6\pi\ \text{cm}^3$ or $391.38\ \text{cm}^3$

Part 5 Review Test

Problem Number:	Problem Answer:	Have Difficulty? See Page:
1.	28/27	15–4
2.	1/2	15–7
3.	1	15–11
4.	128/49	15–7
5.	5/3	15–11
6.	$7^2 = 49$	16–9
7.	$(36)^3 = 46\ 656$	16–25
8.	6.03×10^5	16–31
9.	0.04	16–37
10.	3	16–49
11.	4	16–53
12.	circumference	17–21
13.	radius	17–21
14.	7 in.	17–4
15.	26.138 cm	17–21
16.	$54.367\ \text{cm}^2$	17–21
17.	$24.138\ \text{in.}^3$	17–25

Part 5 Review Test, *Cont.*

Problem Number:	Problem Answer:	Have Difficulty? See Page:
18.	50.113 in.^{2}	17–25
19.	87.019 cm^3	17–33
20.	94.964 cm^2	17–33

Chapter 18   Averages and Percent

1.   4 children/family

2.   3 children/family

3.   2 children/family

4.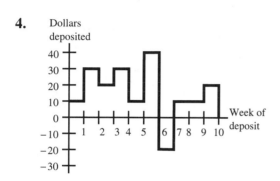

5. $16/week

6.   27.69%

7.   13.85%

8.   37.69%

9.   20.77%

10.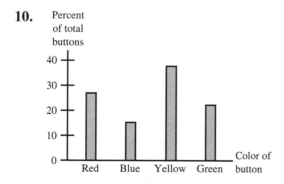

11. $84

12.   28°F/day

13.   28°F/day

14.   28°F/day

Chapter 18 Averages and Percent, *Cont.*

15.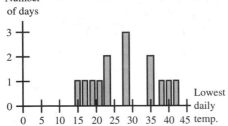

16. 14.29%

17. 42.86%

18.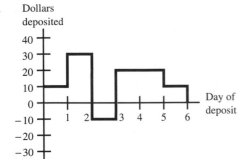

19. $13.33/day

20. $350/week

Chapter 19 Slopes and Rates

1. -3
2. 0; horizontal
3. 4
4. undefined; vertical
5. 3.75%
6. -2.5%
7. $(40)^{1/2} = 6.3246$
8. $(68)^{1/2} = 8.2462$
9. $(137.02 \text{ cm}^2)^{1/2} = 11.706 \text{ cm}$
10. $(56 \text{ ft}^2)^{1/2} = 7.4833 \text{ ft}$
11. 3.5 mi/hr west
12. 2 m/s east
13. 180 miles

Chapter 19 Slopes and Rates, *Cont.*

14.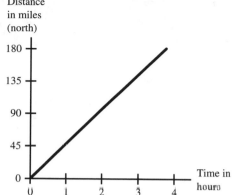

15.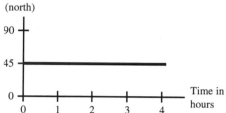

16. 3.5 mi/hr east

17.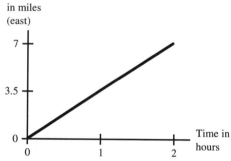

18.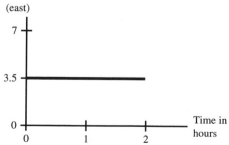

19. 80 m/s south

20. 400 m

Part 6 Review Test

Problem Number:	Problem Answer:	Have Difficulty? See Page:
1.	5 people/family	18–4
2.	4 people/family	18–4
3.	12.5% red buttons	18–9
4.	23.44% green buttons; 21.88% blue buttons; 42.19% black buttons	18–13

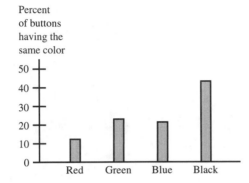

5.	$96	18–17
6.	15.25 students/class	18–4
7.	10 students/class	18–4
8.	10 students/class	18–5

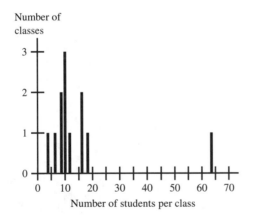

9.	25%	18–9, 18–10
10.	$19/week	18–17, 18–18

Part 6 Review Test, *Cont.*

Problem Number:	Problem Answer:	Have Difficulty? See Page:
11.	Slope = 9/5	19–2, 19–5, 19–9

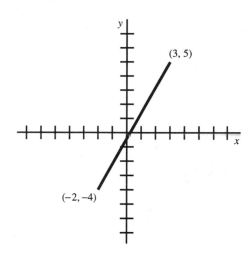

| **12.** | Slope = −0.7 | 19–2, 19–5, 19–9 |

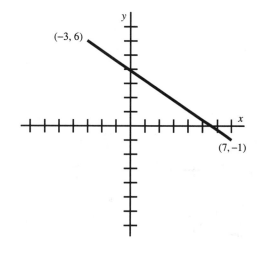

Problem Number:	Problem Answer:	Have Difficulty? See Page:
13.	Distance = 10	19–17

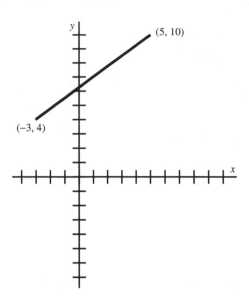

14.	Slope = 2/7	19–2, 19–5, 19–9

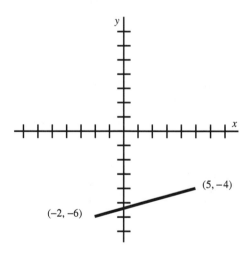

15.	Distance = 7.2801	19–17

Part 6 Review Test, *Cont.*

Problem Number:	Problem Answer:	Have Difficulty? See Page:
16.	Slope = −11/7	19–2, 19–5, 19–9

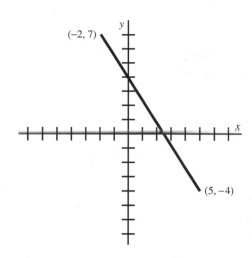

17.	45 mi	19–27
18.	135 mi	19–27

19.

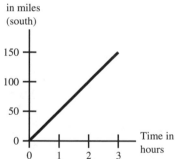

19–28, 19–29

20.

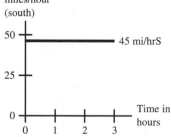

19–28, 19–29

Chapter 20   Sequences and Series

1. 16   **2.** 80
3. 27   **4.** finite sequence
5. infinite sequence   **6.** 6

Chapter 20   Sequences and Series, *Cont.*

7.	$\Sigma = 60$	8.	6
9.	$\Sigma = 258$	10.	$a_3 = 2 \times 3 + 1 = 7$
11.	220 Hz	12.	1760 Hz
13.	55 Hz	14.	$12.25
15.	$14.03	16.	$3.11
17.	$6.06	18.	$8103.27
19.	$9115.07	20.	$34 410.23

Chapter 21   Circular Measure

1.	3π or 9.425 in.	2.	10.5π or 32.99 mm
3.	3 radians	4.	171.89°
5.	$7\pi/6$ or 3.665 radians	6.	third quadrant
7.	120°	8.	second quadrant
9.	270°	10.	$3\pi/2$ or 4.712 radians
11.	180°	12.	π or 3.142 radians
13.	$11\pi/3$ or 11.519 radians	14.	fourth quadrant
15.	1.75 revolutions	16.	−25° or 335°
17.	40.50 m	18.	21.808 m
19.	14.95 in.	20.	11.94 in.

Part 7 Review Test

Problem Number:	Problem Answer:	Have Difficulty? See Page:
1.	37	20–4, 20–12
2.	43	20–4, 20–12
3.	6	20–12
4.	$\Sigma = -6$	20–12
5.	4	20–17
6.	15, 21	20–4, 20–12
7.	3	20–17
8.	36, 108, 324	20–17
9.	2.65 or 265%	20–32
10.	15 years	20–32
11.	10π or 31.42 cm	21–2, 21–3
12.	$5\pi/6$ radians	21–10
13.	270°	21–11
14.	144°	21–11
15.	900°	21–15

Part 7 Review Test, *Cont.*

Problem Number:	Problem Answer:	Have Difficulty? See Page:
16.	180°	21–15
17.	210°	21–15
18.	$5\pi/4$ radians	21–11
19.	390°	21–11, 21–15
20.	30°	21–11, 21–15

Note: Chapters 22 and 23 do not have a Chapter Test.

Chapter 24   Numbers with Letters

1. $-3mn$
2. $4pqr + 9qrs$
3. $-5xy + 68xyz$
4. $31ab + 16bc - 15$
5. $-14rs - 24rst$
6. $x = -3$
7. $y = 4$
8. $z = 11$
9. $t = 5/2$ or 2.5
10. $k = 2$
11. $p = -4/3$ or $-1.3\ldots$
12. $v = 2/5$ or 0.4
13. $d = -21$
14. $g = -2.65$
15. $h = -6/7$
16. $s = 31.3$ cm
17. $v = 13$ ft/sec south
18. $L = 6.1$ in.
19. hypotenuse $= 157.44$ ft
20. $d = 80$

Part 8 Review Test

Problem Number:	Problem Answer:	Have Difficulty? See Page:
1.	30 km/h north	18–4, 22–39
2.	4 h	19–30, 22–39
3.		19–30, 22–39

Part 8 Review Test, *Cont.*

Problem Number:	Problem Answer:	Have Difficulty? See Page:
4. and 5.		18–9, 22–10, 22–31, 22–35

Person	Salary per Year	Age in Years
Jose	$31 460	42
Carmen	$26 000	28
Maria	$21 078.20	21

Problem Number:	Problem Answer:	Have Difficulty? See Page:
6.	22.75 sq ft	17–9, 22–10
7.	7.0686 sq in.	17–21, 22–19
8.	5.093 sq cm	17–21, 22–19
9.	22.695 sq ft	22–19
10.	5	20–17, 20–19
11.	25	20–17, 20–19
12.	780	20–19
13.	3125	20–17, 20–19
14.	22.784 feet/second	14–7, 14–13, 19–27, 19–30
15.	0.77671 mile	14–7, 14–13, 19–27, 19–30
16.	$a + 5ab$	24–5, 24–6
17.	$9x - 2xy$	24–5, 24–6
18.	$y = -3$	24–13, 24–17
19.	$z = -0.4$	24–13, 24–17
20.	16 seconds	24–21

Book 2 Test

Each problem is worth **2 points.** The scoring is given on page B–15.

Problem Number:	Problem Answer:	Have Difficulty? See Page:
1.	7	15–4
2.	40/7	15–7
3.	35/114	15–11
4.	0.625	10–3, 10–7, Book 1
5.	220.938	10–15, Book 1
6.	205.36	10–11, Book 1
7.	$3^4 \times 5^3$	16–2
8.	$1/4^2 = 0.0625$	16–17
9.	$1/9 = 0.111\ldots$	16–25

Book 2 Test, *Cont.*

Problem Number:	Problem Answer:	Have Difficulty? See Page:
10.	7×10^{-4}	16–31
11.	700 000	16–37
12.	$2^{6/3} = 4$	16–53
13.	$3^{4/4} = 3$	16–53
14.	25.6 in.	17–4
15.	17.48 sq cm	17–9
16.	15.975 sq cm	17–21
17.	10 360 cu in.	17–25, 17–26
18.	2851.4 sq in.	17–25, 17–26
19.	623.61 cu in.	17–33, 17–34
20.	43 years	18–4
21.	36.33 years	22–8
22.	$231 818/house	18–4
23.	mode price = $225 000	18–5

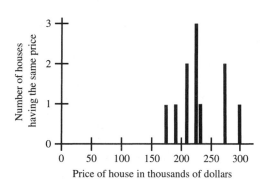

24.	63.64%	18–9
25.	0; horizontal line	19–9, 19–12
26.	−5/9	19–9, 19–12
27.	$(106)^{1/2} = 10.296$	19–17, 19–18
28.		19–29, 19–30, 22–39

Time	First Bike	Second Bike
0 h	0 km	120 km
1 h	25 km	90 km
2 h	50 km	60 km
3 h	75 km	30 km
4 h	100 km	0 km

Book 2 Test, *Cont.*

Problem Number:	Problem Answer:	Have Difficulty? See Page:
29.		19–29, 22–39

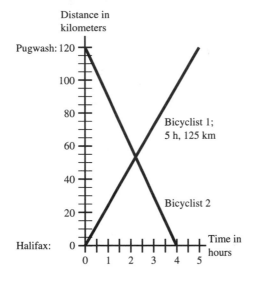

30.	2.25 h approximately	22–39
31.	4.75 h approximately	22–39
32.	4 h	22–39
33.	28	20–3
34.	552	20–3, 20–4
35.	9	20–17
36.	7	20–17
37.	315°	21–11, 21–15
38.	fourth quadrant	21–7
39.	−59π/18 radians or +13π/18 radians	21–10, 21–15
40.	second quadrant	21–7
41.	276.27 ft	21–19
42.	20 miles/hour	19–27, 19–30
43.	29.33 feet/second	14–7, 14–13, 19–27, 19–30
44.		18–9, 22–10, 22–31, 22–35

Person	Age in Years	Salary in Dollars/Year
Jane	29	$24 000
Alice	22	30 000
Karl	19	17 700

Book 2 Test, *Cont.*

Problem Number:	Problem Answer:	Have Difficulty? See Page:
45.		18–9, 22–10, 22–31, 22–35

Person	Age Now (Present Age) in Years	Age in 5 Years	Age in 10 Years
Maria	25	30	35
Jose	55	60	65
Carmen	13	18	23

Problem Number:	Problem Answer:	Have Difficulty? See Page:
46.	$27xy + 29xz - 5$	24–5, 24–6
47.	$p = 2.5$	24–13, 24–17
48.	$t = 0.6$	24–13, 24–17
49.	$y = 13$	24–21
50.	$t = 6$ seconds	24–21

Book 2 Test Scoring

100 points maximum score
95–100 Excellent
85–94   Good
75–84   Fair
65–74   Poor
0–64   Reread Appendices D and E;
reread the referenced parts of this book;
then take the test again.

Overcoming Anxiety

At one time or another, we all have anxious feelings. The unknown, and those things known or believed to be "scary," make us anxious.

A parent, teacher, or friend may say something that adds to our own anxious feelings regarding math. For example:

- Math is tough.
- You have to be smart to do math.
- There is only one answer to a math problem.
- You can't bluff your way through math.

When we watch other people "suffer" as they work their way through math, then we become even more anxious.

What *is* anxiety?

Anxiety summarizes in one word some very complicated feelings. We feel uneasy, "anxious." Anxious feelings can occur when we have to face bodily changes (an operation) or lifestyle changes (financial, births and deaths, or retirement). We have fewer feelings of anxiety when we are in good physical condition and feel that we are in control of our life.

We become less anxious and feel more secure when we

- exercise daily;
- relax by meditating or deep breathing; or
- talk with someone who feels more secure.

In the extreme, we may need to discuss the problem with our physician or, if necessary, a therapist.

Most people can work through their anxieties by

1. wanting to conquer the related fears, and
2. doing something positive about them.

Therefore, we have developed two simple sets of steps. They are presented in two of our appendices:

- The first set of steps are for those who have difficulty remembering technical facts (Appendix D).
- The second set of steps are for those who have difficulty taking tests (Appendix E).

Read Appendices D and E. Try this different approach.

Learn the simple steps for controlling your anxieties. Plan your life and time better. (This will help you reduce the stress caused by anxiety.) As we live our lives, we may have to walk some unfamiliar roads. This book, and especially Appendices D and E, are "road maps" for technical subjects such as math and science.

We have split the math topics in this book into many small parts. Each part can be learned—a few facts at a time. Then, as you gradually piece these facts together, the parts will begin to make sense. The learning of additional math and other technical material becomes much easier once you know and understand the fundamentals.

Is anxiety a problem for you? Your fears should soon diminish.

Why?

Because you have already taken the first step: You opened the pages of this book!

Memory Methods

When we like something, we usually remember it. an anniversary, a birthday, the end of a war. Math and other technical subjects can be unpleasant for us. Therefore, we must work harder at remembering.

Math and other technical subjects can be practiced and learned both in a group setting (such as a classroom) and on your own. Let us look at both situations.

The Classroom (Group)

Try the **LARD** approach:

- **L**isten attentively—we all daydream from time to time. Force yourself to concentrate on the teacher, the chalkboard, or another student who may be talking.
- **A**sk questions—don't be shy. If you knew it all, you wouldn't be there. You are practicing **communication** when you ask a question.
- **R**equest help—if you cannot learn it on your own, then request individual assistance. Request help from your tutor, your teacher(s), or another student. When other students help you, they learn their math faster.
- **D**escribe your understanding—when you explain your newly learned math topic, two things can happen. First, you hear yourself say it and that is **reinforcement.** Second, you can be told if your wording is right or wrong.

The **LARD** approach—**L**isten, **A**sk, **R**equest, **D**escribe—is an excellent classroom method for learning math.

On Your Own

When are the best times for you to study math?

When you first awake and you are alone?

During a break at work?

While waiting in traffic that is going nowhere?

While eating a brief meal?

In the evening after a brief rest?

Some other set-aside time?

Only you can select from, and add to, this list.

Set aside from 15 to 90 minutes; have your math book with you and ready. Plan that time before it arrives; defend it from all distraction.

Your study environment is important:

- Find a quiet place that has good lighting.
- Remove any item that may distract you. Remember: You may secretly want to be distracted from studying. **Fight that thought.** Concentrate on your studying. Get it done and go on to more pleasant things!
- Make yourself comfortable. Choose a surface on which to work and a chair that fits you well.
- Avoid eating while studying if it distracts you. Concentrate on your studying. Snacks are an excuse to avoid the studying. (A cup of soup or a sandwich is all right if that is the end of your eating.)
- Study only when you are rested. A twenty to forty minute nap, followed by an hour (more or less) of studying, is known to be very efficient. You will learn more than you could during three hours of trying to study when you are tired.
- A well-lighted place to study that has no distractions is known to be both effective and efficient.

There are ten **study commandments** that have proved very successful for other students. Try them; you might like all or some of them!

Study Commandments

I   *Scan the chapter once.*

Note the concepts or procedures that seem important to you. Use a colored pen or pencil to identify them.

II   *Work the examples.*

If you find you have difficulty with an example, then review the related material. Do not dwell on one example; go on to another one. Come back to the troublemaker later.

III   *Read the summary.*

The summary contains information that the author(s) believe you will find important. You may find other material in the chapter that is also important to you. Add your other information to the summary.

Study Commandments

IV   *Prepare 3 × 5 cards.*

Use either 3 × 5 cards or small pieces of standard-size paper to note the important concepts and procedures. Write one question on the front. Write its answer on the back. This is your flexible study guide. For example:

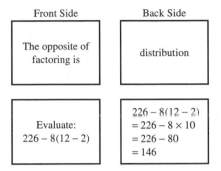

Front Side

The opposite of factoring is

Back Side

distribution

Evaluate:
$226 - 8(12 - 2)$

$226 - 8(12 - 2)$
$= 226 - 8 \times 10$
$= 226 - 80$
$= 146$

V   *Take the chapter test.*

Each chapter contains a test. You should follow the test instructions as indicated there. Determine the answers. Make a record of how long it took you to take the test.

VI   *Mark your mistakes.*

Determine your test score by comparing your answers with those in Appendix B. More important to you, mark those problems that were wrong.

VII   *Reread portions of the chapter.*

Return to the chapter; reread the portions that gave you difficulty. Revise, and add to, your 3 × 5 cards if necessary.

VIII   *Correct your mistakes.*

Be certain you can now work each problem that gave you difficulty. Work the problems over again if you find that to be necessary.

IX   *Review your 3 × 5 cards.*

Technical material is best learned in small, frequent doses. Review your 3 × 5 cards at least daily at first, then less frequently. Always review them before you take a test.

X   *Sort your 3 × 5 cards.*

Make two piles. Make one pile of those cards whose answers you know well. Study the other pile three or four times daily until you know them well also.

SUCCESS!

Finished? Great! Remember: Math topics usually build upon each other. You will feel much better about math topics if you practice, practice, practice!

Test Taking

The taking of tests can be made easy, or it can be made difficult. You can make it easy for yourself by following the guidance given in Appendix D. If you know your facts well (and have practiced, practiced, practiced), then you should do well on most tests.

Before taking a test—relax! Oh, well—at least *try* to relax. Most of the time, the following ideas will work for you:

1. Work, or attempt to work, the test problems starting with Problem 1.
2. If you cannot recall how to work a problem, then *mark that problem* and come back to it later.
3. Work the last test problem; then take a few deep breaths (to relax you).
4. Go back to those *marked problems* that caused you difficulty; try again to work them.
5. Do *not* change an answer because you are nervous; have a good reason for change.

Once you have completed these five steps, stop! You have done enough and deserve a reward. Reward yourself!

The Prime Numbers: 1 through 500

1	53	131	223	311	409
2	59	137	227	313	419
3	61	139	229	317	421
5	67	149	233	331	431
7	71	151	239	337	433
11	73	157	241	347	439
13	79	163	251	349	443
17	83	167	257	353	449
19	89	173	263	359	457
23	97	179	269	367	461
29	101	181	271	373	463
31	103	191	277	379	467
37	107	193	281	383	479
41	109	197	283	389	487
43	113	199	293	397	491
47	127	211	307	401	499

G

Calculator Key Options

Hand-held calculators are almost a necessity today. They provide us with the opportunity to perform calculations more rapidly and more accurately.

We recommend that you purchase a simple calculator. Therefore, we recommend the following two groups of keys or functions:

Necessary Keys

- the ten decimal symbols (0 through 9)
- the decimal marker (.)
- the operations $+$, $-$, $\times$, $\div$, $=$
- the grouping symbols ( )
- y^x and its inverse (the xth root of y)
- x^2 and the square root of x
- exponential for scientific notation (EE or EXP)
- reverse sign ($+/-$)
- sine (SIN) and its inverse*
- cosine (COS) and its inverse*
- tangent (TAN) and its inverse*
- $\log_{10}$ and inv log (LOG_x and 10^x)
- ln and inv ln (LNx and e^x)
- RESET or CLEAR

---

*The sine inverse operation may be noted as $\sin^{-1}$, arcsin, or invsin. It is spoken, **"the angle whose sine is."** The same is also true for the cosine and tangent trigonometric functions.

Optional Keys

- percent (%)
- reciprocal (1/x)
- *x* factorial (x!)
- degrees/radians/grads (DRG) for Chapter 21
- π for Chapters 17 and 21
- one or more memory locations

We further recommend that your calculator be solar powered. Then, during a test, there is no battery to fail.

How do you test a calculator? Enter "4 + 3 +" and then press the "=" key several times. If it stays at "7," then purchase it. Otherwise, go to another model.

Index to Book 1 and Book 2

Chapter references that start with 1 through 14 are in Book 1.
Chapter references that start with 15 through 24 are in Book 2.
Numbers in **bold** refer to definitions in the chapter glossaries.

Axis, reference. *See* Reference axis

Bar. *See* Scalar. *See also* Bar display
Bar chart (or graph), 10–23, 18–5, **18–22**
Bar display, 2–2, 3–2, 11–7, 11–29, 11–35
Base (of a surface). *See* Parallelogram; Rectangle; Square; Triangle
Base (of an exponent), 16–5, **16–59, 17–43,** 20–23
Base ten. *See* Base (of an exponent)
Binary system, 1–27
Binomial, 24–5, **24–26**
Borrow, 3–7, 3–8, 3–11, 3–15
Borrowing numbers
  for decimal numbers, 10–11
  for subtraction, 3–7, 3–8
Braces, **7–30.** *See also* Grouping symbols
Brackets, **7–30.** *See also* Grouping symbols

Cancellation
  of numbers, 9–19, 9–20, 9–21
  of units. *See* Unity method, units that cancel
Carry, 2–12, 3–15
Carrying numbers
  for addition, 2–11
  for decimal numbers, 10–11, 10–15
  in multiplication, 4–13
Cartesian coordinate system. *See* coordinate system, cartesian
Celsius, **14–38**
Center (of a circle), 17–3, **17–43**
Change. *See* Delta. *See also* Rate; Slope
Change in angle. *See* Delta theta
Chart, 19–30
  addition. *See* Addition, chart
  bar. *See* Bar chart
  pie. *See* Pie chart
Check. *See* Self-checking
Circle, 17–3, 17–21, 17–41, **17–43,** 21–1
Circular measure conversions, 21–9, 21–10

Circumference, 17–21, 17–41, **17–43.** *See also* Perimeter
Clockwise (CW) angle, 21–3, 21–27, **21–28**
Closed surface. *See* Surface, closed
Coefficient, 24–1, 24–2, 24–25, **24–26**
Column names
  decimal. *See* Decimal column names
  fraction. *See* Decimal column names
  hundreds. *See* Hundreds (column)
  hundredths. *See* Hundredths (column)
  number. *See* Number column names
  tens. *See* Tens (column)
  tenths. *See* Tenths (column)
  thousands. *See* Thousands (column)
  thousandths. *See* Thousandths (column)
  units. *See* Units column
Combinations, 6–3, **6–17**
Combined math operations (signed numbers). *See* Signed numbers, combined math operations
Combining fractions. *See* Fractions, combining
Common denominator (least), 9–4, 9–6, 9–30, 9–37, 9–61, **9–62, 9–63.** *See also* Equal denominators
Common difference. *See* Arithmetic progression
Common factor, 8–1
  canceling. *See* Factor, canceling
  largest. *See* Factor, largest common
  reducing. *See* Fractions, reducing
Common ratio. *See* Geometric progression
Commutative law, 7–19, **7–30**
  of addition. *See* Addition, commutative law
  of multiplication. *See* Multiplication, commutative law
Comparing fractions. *See* Fractions, comparing

Comparison. *See* Number comparisons
Complex fraction. *See* Fraction, complex
Complicated fraction. *See* Fraction, complex
Composite number(s), 6–7, 6–8, 6–17, **6–18**
Consecutive numbers, 1–6, **1–28**
Consecutive order. *See* Consecutive numbers
Conversion
  circular measure. *See* Circular measure, conversions
  equivalents of measure. *See* Measure, conversion equivalents
Converting decimal fractions to decimal numbers. *See* Decimal fraction, converting to decimal number
Coordinate system, cartesian, 19–9, 19–23, 19–46, **19–48**
Cosecant, 21–20
Cosine (of an angle), **21–28**
  formula, 21–20, 21–22
Cotangent, 21–20
Counterclockwise (CCW) angle, 21–2, 21–3, 21–27, **21–28**
Counting numbers, 1–5, 1–6, 1–28, **1–29,** 6–17. *See also* Natural numbers
Cube root. *See* Root, cube
Cube (surface area and volume), 17–25, **17–44**
Cubing a number, 16–50
Curve, 17–3, 17–41, **17–44,** 21–1
  of mean average. *See* Average, mean, curve of
Curved surface. *See* Surface, curved

Decade, 20–23, **20–38**
Deceleration, 19–41, **19–48**
Decimal, 1–5
  column names, 10–24. *See also* Number column names
  digit. *See* Digit, decimal
  fraction. *See* Fraction, decimal
  column names, 10–24
  converting to decimal number, 10–2